Andrian

&

BITCOIN

ARE THE UNSTOPPABLE FORCE !

Change is constant &

with it comes opportunities!

Success for 2020

The Tedster

BITCOIN
THE UNSTOPPABLE FORCE

The Next Generation Internet

TED HILL

Bitcoin: The Unstoppable Force

By Ted Hill

ISBN: 978-1-7332041-1-8
Book Design & Publishing done by:
Susmita Dutta
Global Book Publishing
www.globalbookpublishing.com

Disclaimer: The Publisher and the Authors make no representations or warranties with respect to the accuracy and completeness of this work and specially disclaim all warranties, including without limitation warranties of fitness for a particular purpose. No warranty may be created or extended by sales or promotional materials. The advice and strategies contain herein may not be suitable for every situation. This work is sold with the understanding that the Publisher is not engaged in rendering legal, accounting, or other professional services. If professional assistance is required, the services of a competent professional person should be sought. Neither the Publisher nor the Author shall be liable for the damages arising here from. Then fact that an organization or website is referred to in this work as a citation or potential source of further information does not mean that the Author or the Publisher endorses the information the organization or the website may provide or recommendations it may make. Further, readers should be aware that internet websites listed in this work may have changed or disappeared between when this work was written and when it is used.

This book is dedicated to my Mom (known as Gran to all) who was passionate about technology. She was an avid reader, part of many book clubs, and she also volunteered for 16 years, at a youth drop-in centre interacting and guiding with teens. She was always intrigued by what's next in technology starting from her computer or Facebook or Youtube, and wanting the latest iphone or tablet. I spoke to Gran about writing a book on Bitcoin. Her eyes sparkled and wanted to know how she could get some. Gran and I shared many interests about sci-fi movies like Star Trek - the voyages "where no man has gone before". I told her that I would put that in my book. What I saw and what was inspiring to me, was at her age still being excited about what is next, excited of what all the tomorrows would bring.

CONTENTS

INTRODUCTION

The technological genie is out of the bottle! Multiple innovation platforms are evolving at the same time.

What more is possible?

We are still in the earliest stages of true Fintech, as the future impact of cloud computing, the Internet of Things, artificial intelligence, and blockchain cannot yet be estimated. Each year, tech companies are digging deeper into the financial services value chain and creating new market structures in underbanked developing countries.

In robotics, Amazon seems to be the front runner, having increased their one thousand warehouse robots in the year 2013 to over 200k in 2019. The industrial robot is already a commonplace in industries like automobiles and manufacturing. Now, collaborative robotics (cobots) can be trained to enter every industry, leveraging human effort.

Crispr gene editing now gives us the ability to correct the programming errors that result in disease. Gene editing kits can now be purchased individually with little regulation in place.

DNA sequencing is bringing real science into healthcare, by breaking down the three billion pieces of code in our DNA.

Bitcoin, Blockchain, and fifteen hundred plus virtual currencies have taken center stage at the moment.

Bitcoin is just a clever technology that people have attached some sort of value to, trying to turn it into money. It won't work. It's a 'bubble'. It's full of security issues and will never scale. Everyone is getting hacked, and its anonymous nature is being used by the bad guys.

That is pretty much what was said about the internet. Wrong then and wrong now!

This book is meant to create a new awareness and an ability to have a reasonable, intelligent conversation about the transformation of money and how we got to Bitcoin and Blockchain. It's for baby boomers, millennials, and Generation X, that have heard about the internet's next era but only know the negative reports about it and very little of the benefits its impact will have.

The world is built on numbers. There's something very mathematical about our universe, and the more carefully we look, the more math we seem to find. Mathematical concepts are not wired into the human condition. They are learned, acquired through cultural and linguistic transmission. With everything physical there is an algorithm or mathematical formula behind how it was made, and this is how Bitcoin came to be. Now we are fully immersed in the digital world.

So, digital money is coming. No, it is already here and has been for decades. About 97% of all financial transactions are already digital, and the rest is in the form of the paper we throw around called cash. This is what most of us know as money. Our school systems in the majority of countries around the world do not teach what money is and how it actually has value. We are taught to get a good education so we can work to earn money. It is a system set up to keep the masses in the dark.

What was the impact of the internet on the world and what did it leave out? It certainly gave the power to the free flow of information without restrictions, but it left one-third of the world's population behind. What do most of us really know about the different types of internet? The Dark Net, the Deep Web, the Internet of War, the Internet of Things.

Virtual currencies mark a paradigm shift in how we think about payments, traditional financial processes, and engaging in economic activity. It will bring in higher levels of trust, responsibility, and accountability for all people, organizations and governments.

The internet did bring the infrastructure for Bitcoin and Blockchain to work. The foundation does not need to be built and is now going to have the greatest invention we have seen built on top of it. Just like the web and email services are technologies in themselves built on the internet, this new layer of technology will revolutionize and disrupt long-standing institutions, creating an enormous change in businesses' ability to trust and act with transparency—perhaps help restructure how societies should be or become.

It is moving at warp speed, so the attempt here to provide the basics, the foundation of what it is and what is to become the next phase of digital money. The Digital F.A.N.G.—Facebook, Amazon, Netflix, and Google—were the four major start-ups that have already captured the opportunities in the digital world and advanced them even more. Uber and AirBnB have created new co-ordinated economies that will tap into blockchain application, and that is just the beginning. The Lightning Network is already moving bandwidth on the net at breakneck speeds which promise the new technologies will have no latency.

This book flows through exponential technology, store of value, and the new coordinated economies. How did the banks become so powerful? There are a billion people today who no longer use cash, and world government wants cash to disappear altogether.

So, what is Bitcoin and the underlying technology that makes it work? What problems does it solve and why do you really need to know? Is it really going to impact you? Has the Blockchain in printed form actually been used for centuries to before taking its modern form? More importantly, how can a better understanding of historical changes indicate what's to come?

Are the Blockchain applications capable of impacting climate change? The problems that Bitcoin and blockchain will solve are enormous and billions have already been invested by governments and big business around the globe, but what are the real-world applications in play today?

The following chapters go through the impact of technology disruption, but chiefly address the next generation's technology. Types of digital payment methods available today and the amount of digital currency in global play –will be covered, along with detailed descriptions of what and how Bitcoin and Blockchain work. Terminology built around the new surrounding technology will be explained, and we'll look at why there are so many cryptocurrencies in 'circulation'.

Manage your entry to the new money. Understanding the basics of how to buy cryptocurrencies, how to store them, and even where to spend them is vital. At the end of the day, or the end of the decade, Bitcoin and Blockchain will be everywhere, so let's move beyond the sense of 'magic', saying "It just works"!

1

THE TECHNOLOGY DISRUPTION

In 2019 Bitcoin turned ten years old. The first lines of code were committed to the bitcoin blockchain on January 3rd, 2009, a few months after the publication of the original whitepaper. These lines of code, referred to as the "genesis block," are credited to the person or persons known as Satoshi Nakamoto.

But Nakamoto vanished. The original Satoshi Nakamoto white paper begins, "A purely peer-to-peer version of electronic cash would allow online payments to be sent directly from one party to another without going through a financial institution.

Technology Disruption

Bitcoin and Blockchain are leading the way in the next generation of the Internet!

Is Bitcoin a purely speculative opportunity as a bubble waiting to burst. What is the issue Bitcoin will solve? Shouldn't governments and banks put regulation in place before cryptocurrencies take over? Everyone is getting hacked. How can this be money in question? Don't we already have digital money?

There are so many more questions than answers on the path of the “new money”.

The convergence of technology is accelerating disruption—everywhere!

Exponential technologies are colliding into each other, reinventing products, services and industries. Technologic advancement is just like this, but can we keep up? Can we adapt as quickly as technology is changing?

Autonomous vehicles were some kind of “Jetsons” idea—the concept of flying cars that would take fifty years to play out. Well, driverless cars have arrived and within ten to fifteen years their operative ability will be fine-tuned, as Artificial Intelligence provides the exponential progression. But there is more to that and how it is all happening so fast.

The Blockchain and AI are definitely near the point where they will soon be crossing over, merging and leveraging each other’s technologies, scaling faster than they would on their own path. The creation of so many new applications and opportunities are creating a technological boom.

There’s good reason for that. The convergence of technology is accelerating disruption—everywhere!

Standard Progress is like going for a walk. Taking one step forward, you advance one step. You take ten steps, and you’re ten steps away from where you started.

Exponential Progress is the kind taking place in technology labs that radically changes the equation and radically accelerates the pace of change we see in the world. With exponential progress, each step is DOUBLE the step before it. You take one step. Now double that . . . and your second step is the equivalent of TWO regular steps. By the time you get to your tenth step, the distance travelled is equivalent to 512 regular steps! By the time you get to the twentieth step, it’s like you’ve taken 524,288 of them!

The Tipping Point

Exponential progress is progress that ‘snowballs’.

The first stages of exponential growth aren’t incredible. As a result, disruptive innovations—like self-driving cars—may strike us as nothing more than a fantasy in when they’re only twenty years away. But computing power and Artificial Intelligence is advancing at incredible—and accelerating—rates. At the same time, the cost of computing power has plummeted.

Computing Beyond Moore’s Law

In 1965, Intel co-founder Gordon Moore observed that the number of transistors per square inch on a microchip had doubled every year while the costs were cut in half. It simply meant that computers and computing power would both get smaller and faster over time.

However, Moore’s law is slowing down—at least for our classical computers. His prediction proved accurate for several decades, until reaching a saturation point around 2015, when the pace slowed, due to shifting of technologies and advanced breakthroughs.

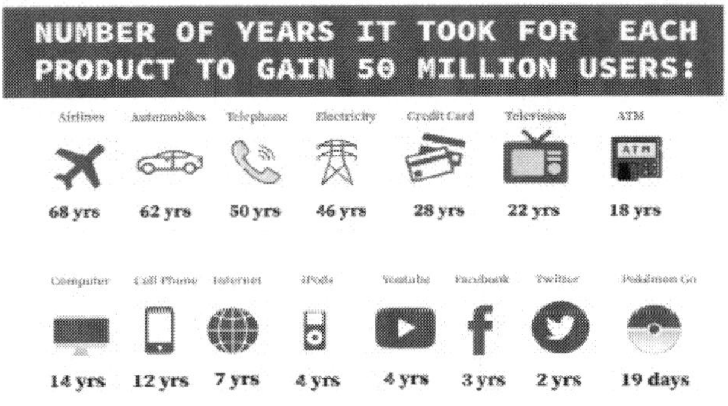

The Wright Brothers solved an unsolvable problem with flight.

Satoshi Nakamoto, the unknown Bitcoin inventor, made it possible to securely send money anywhere in the world without banks or any other intermediaries with Bitcoin and Blockchain.

Whenever the unsolvable happens, it inspires incredible innovation.

When the Wright Brothers launched and flew their first plane, Kitty Hawk, naysayers said it would never hold more than two people, that it was dangerous, and had no value.

The legacy it left behind was months-long crossings of the Atlantic Ocean down to just a few hours. Just sixty-six years later, men would fly to the moon. Today, there are 500k people in the air daily around the globe.

And now, in 2019, Kitty Hawk has been resurrected. Behold the emergence of the futuristic Kitty Hawk. The 'electric flight' start-up, backed by Google co-founder Larry Page, unveiled its third aircraft which is a third of the size of a Cessna, does vertical takeoffs, can travel fifty-five miles in fifteen minutes, and is roughly one hundred times quieter than a helicopter. What more is possible? NASA has their first all-electric, experimental aircraft plane that can hit a cruising speed of 172 mph at eight thousand feet.

Innovative Ecosystems

"Even if you are on the right track, you'll get run over if you just sit there." —Will Rogers (Humorist)

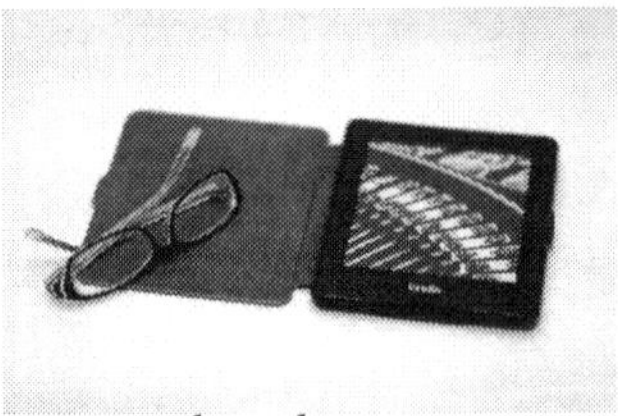

Amazon has almost no stores

Uber owns almost no cars

Facebook creates no content

Alibaba has no inventory

AirBnB owns no real estate

Bitcoin owns no dollars and needs no banks

It took early humans over a million years after discovering fire to invent a wheel. Then, it took only two thousand years to invent a printing press, and about two hundred more to build a telescope. In the two hundred years that followed, we jumped from the telescope to a steam locomotive, and two hundred years after that we built a space shuttle. Then, it only took two more decades for us to start modifying our DNA.

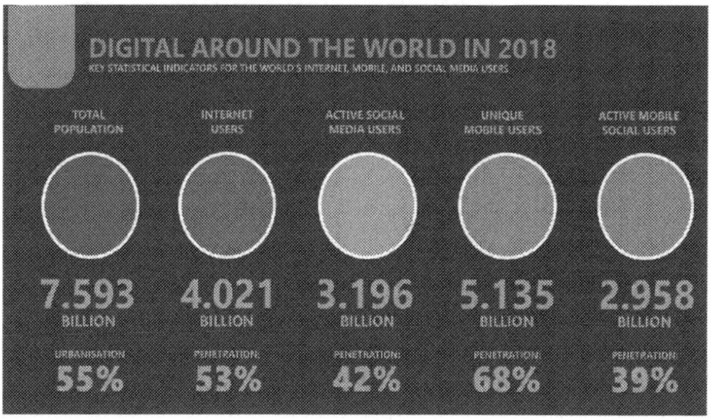

In less than 20 years, billions of people are now connected to the Internet worldwide; shopping online and already making cross border payments. Imagine what the next twenty years may bring when the convergence of these technologies become symbiotic, along with so many new, powerful applications. What more is possible!

Ok—let's keep it simple

Everyone understands technology is changing at a rate that can't be matched by human intellect. We have seen the future with Star Trek since 1969 and Orwell's 1984, including the advancement of Artificial Intelligence. This is just another movement in technology—but at an exponential rate.

We have a world that has been built on many viewing platforms, all of which impressed our values and historical perspectives. But disruption and transformation continue, especially for rock-solid companies that must adapt or perish. The management firm Innosight which performs global strategy consulting, released a 2016 report titled "Corporate Longevity: Turbulence Ahead for Large Organizations", reporting upon the 500 largest U.S. publicly traded companies (5.5 billion a year in revenue to be in the Top 500).

"In 1990, average tenure in the top 500 had narrowed to 20 years and is now forecast to shrink to 14 years by 2026. At the current churn rate, about half of today's 500 firms will be replaced over the next 10 years as we enter a period of heightened volatility for leading companies across a range of industries. The constant turnover in the Fortune 500 is a positive sign of the dynamism and innovation that characterizes a vibrant consumer-oriented market economy, and that dynamic turnover is speeding up in today's hyper-competitive global economy."

This all points to the disruption from technology.

So why wouldn't Bitcoin and Blockchain be the next major changing event? Or is it just another bubble or the biggest bubble ever? Could it be the greatest transformation of the internet itself?

One way of thinking: today, a rumor that gets started by nine a.m. can circle the globe and could have a major following by 10am.

Today paradigm shifts occur in only a few years time!

By 1000 A.D., progress was almost non-existent, and a paradigm shift required a century or two. In the 19th century, we saw more technological change than in the nine centuries preceding it.

Then, in the first twenty years of the 20th century, we saw more advancement than in all of the 19th century.

The World Wide Web did not exist in anything like its present form just a decade ago, and it didn't exist at all two plus decades before that. TechJury.net reports there are now almost 4.4 billion websites, but only a quarter of them are being used.

As these exponential developments continue, we will begin to unlock unfathomably productive capabilities and begin to understand how to solve the world's most challenging problems. Change is constant and moving full speed ahead. Speaking strictly of innovations, there are over 500k patents filed every year.

It doesn't get any better than this!

Does disruption come in many forms, besides technology? Isn't it really just a dramatic breakthrough innovation in any area of a particular field? Some believe that any technology-based company offering a hot, new product is actually practicing disruptive innovation.

Are all innovations disruptive by nature? Or does an innovation have to be a breakthrough to be disruptive? Or is disruption only related to technology?

Disruptive transformational innovation is an entirely different phenomenon that utterly disregards all of the above-mentioned misconceptions.

Bitcoin & Blockchain will transform money the way the Internet transformed commerce—and that's just the beginning! Riding along the new wave of the greatest technology advancement will impact everyone and every business in years to come! We have a global economy that is built on centralization and the power within. A redistribution of how the power works in economies is here and more is coming. The changes will be profound.

What more and what else is possible?

Bitcoin, an invention created as an open sourced project in 2009 by mystery developer Satoshi Nakamoto, was for six years dismissed for the same reasons the internet had been.

In 1992, internet conferences held little value. People had little vison of what the internet was to become. They said it couldn't scale and was used mostly by drug dealers. Most people were unable to understand what the problem was that the internet would solve.

The internet, with the web as a new killer app, was bringing together unrelated buyers and sellers in seamless, low-cost ways. Around the world, new business models began in business and economics. The majority were simply people with ideas. Many dot-com business plans were predicated on the assumption that by using the internet, they would bypass the distribution channels of existing businesses and therefore not have to compete with them (when established businesses with strong existing brands developed their own internet presence).

The dot-com bubble burst in March, 2000, and by 2001, the majority of previously involved parties had ceased trading. Dot.com became Dot.Gone in the ecommerce space, bursting like a true bubble. But driven by commerce, ever greater amounts of online information, knowledge, and social networking boomed.

No one would have ever imagined that businesses could be run across the globe through a computer. All over the world, online education has brought a revolution in the field of education and knowl-edgsharing.

The internet brought inclusion. At least to some. Anyone anywhere with access now held the keys to the world of information, rather than just books, directives from government or big business—a transformation of learning, creating a social evolution that has impacted everything we do.

To look back on some famous predictions is almost comical:

- IBM President Thomas Watson, 1943: "I think there is a world market for maybe five computers."
- Ken Olsen, President of DEC computers, 1977: "There is no reason anyone would want a computer in their home."
- Paul Krugman, Nobel Laureate, Economist, Professor of Economics, 1998: "By 2005 or so, it will become clear that the Internet's impact on the economy has been no greater than the fax machine's."
- Blackberry CEO Thorsten Heins, 2013: "Tablets themselves are not a good business model, and that, in 2018, no one will be buying a tablet of any kind."
- Generation Z person - in 2040: What took so long to go digital?

Research firm IDC shows 120 million people owning a cell phone when Apple launched its first iPhone. Today, over five billion own a Smartphone in a global population of 7.7 billion.
Eighty-three million people are added to the planet each year—there is a market.

Apple has been hitting grand slams for twelve years but is on its tenth version of iPhone, with sales numbers dropping. The increase in price per phone cannot compensate the revenues dropping 12% a year over the last three years.

Apple's R&D group spends one billion a year, but Apple needs to pivot and try to pull off another iPod/iPhone/iPad megahit, or its best days are gone.

Or have they?

Apple's 'wearables' sells more watches than any other watchmaker on Earth and is flush with $180 billion—cash in the bank. They have more than Rolex, Timex, and all the Swiss watch manufacturers combined.

Sony was the Apple of its time, dominating the 90's, as it dominated TV, video games, and portable music players. But Sony ran out of big ideas. Sony was one of many companies releasing their software technology as proprietary meaning it could not interact with other. Open source technology now rules, meaning developers have full access to code to build upon.

The point here is that the marketplace changes rapidly, technology is exponential, so the Amazons and the Apples of the world need to continue to pivot and stay ahead of the curve.

One of Apple's significant pivots has been their move into wearable devices. They are now moving towards and adopting Blockchain and the cryptocurrency, and a recent software release features what Apple calls a "cryptographic" developer tool. Open source technology will allow developers to adjust to whatever the regulation might be.

Transformation—Coordinated Economies

- Uber and Airbnb are powerful transformers creating new economies for all people by the people.
- Bitcoin is the new currency for all people involving trust, proof of work, and opportunity for financial inclusion.
- Money is a language. Money is a store of value. Money is a system of control granted by an authoritative figure.

Say hello to the coordinated economy, a major shift in technology with long-running implications for all.

What is the coordinated economy?

Put simply, the coordinated economy is a modern economic system that leverages technology and incentives to unlock the power of the people. A coordinator, like Uber, takes an industry and basically creates a whole new model.

What are the key characteristics of a coordinator?

- **Big demand market:** The coordinator needs to operate in a market where demand is seemingly bounded only by population.
- **Monopoly supply market:** Monopoly is the form of power structures in which power effectively rests with a small number of people. The coordinator needs to operate in a market that is traditionally characterized by constrained supply relative to the demand, where constrained supply is largely the result of a limited number of suppliers.
- **Consumer pain points:** The market in which the coordinator operates needs to be traditionally seen as one with consumer pain points as a result of a supply-demand imbalance. Those pain points will most often be high prices and low convenience.
- **Power of the people:** The coordinator needs to operate in a market wherein the many can produce the same service or good at comparable or better quality as the few.
- **The right incentives:** The coordinator needs to employ an incentive structure (financial, social, or both) which promotes exceptional output by the service or good covered.
- **Results:** As a result of the coordinator's democratization and coordination efforts, consumer convenience needs to be higher, transaction price points need to be lower, and supplier benefits need to be better.

Why is the coordinated economy better for suppliers? Coordinated economy principles—when applied properly—also generate optimal supplier outcomes. Because the majority of coordinated economy suppliers in a democratized system are new, lower prices from the monopoly norm are not negative (as they are for traditional suppliers). Instead, because these suppliers weren't being paid before democratization allowed them to get paid, getting paid at all is a net positive.

Demand for car rides is largely "unlimited," bounded only by population and vehicle parameters.

At one time, there was one supplier—taxis. The result was that unlimited ride demand was coupled with limited driver supply. That low supply, high demand dynamic led to high ride costs for consumers, and a great deal of consumer inconvenience due to lack of supply.

Uber changed this entire system, leveraging the power of the people by transforming the car transportation industry from a single-supplier ecosystem to anyone with car. And now they have even started supplying cars if you don't have one.

The net result is that supply and demand now largely equal each other in the car transportation industry. Consumer convenience is higher. Transaction price points are lower. People can now get paid for shuttling others around, something that wasn't possible before.

Why Should Investors Care?

Investors should care about the coordinated economy because it is this era's biggest consumer-facing digital trend. Much like the advent of the modern internet unlocking tremendous value in consumer-facing businesses and the social web value in consumer-facing business over the past decade, the coordinated economy will unlock equally tremendous value in consumer-facing businesses over the next decade by substantially improving consumer and supplier outcomes.

Thus, if you're looking for tomorrow's biggest and brightest stars, you need to examine companies that either are or have the potential to be coordinators. Bitcoin and Blockchain will accelerate the power of coordinated economies.

Here are three examples of coordinated economy companies that could explode higher over the next several years.

- **Shopify:** Shopify is democratizing the e-commerce industry by providing commerce tools to anyone and everyone who has something to sell. It's a big demand market. At some point or another, everyone needs to shop. Demand in this market is very big, and it constitutes almost every single individual in the digitally connected world. The right incentives: Shopify simply gives tools to individuals to sell products. Thus, a majority of sales that happen through Shopify websites end up in the seller's pockets, meaning that the incentive structure is in place here to promote exceptional supply-side output.
- **Amazon:** Oligarchic supplier market: Supply in the e-commerce, and more broadly commerce, market is traditionally oligarchic (few suppliers in one market) since there are multiple barriers to selling items, both in-store and online. That is why Amazon controls about 50% of the e-commerce market today.
- **Etsy and eBay:** As community commerce platforms like eBay and Etsy prove, there are thousands, and even millions, of people globally who have the ability to create, market, and sell various products that are of comparable and sometimes higher quality than products found at Walmart or Target.

Let's take a step back and breathe?

There is so much more that will happen in technology to come, and we're on the cusp of doing so. What coordinated economies are already in the making and what more may appear out of nowhere? You could ask the Alexa's, Siri's, and Cortana's in our world, as they will become powerfully sophisticated and portable, actually answering with extreme accuracy with built in AI.

In fact from a healthcare perspective, this group of 'know it alls', may start to become your first medical opinion and your Doctor will be second.

Can the world keep up?

2

THE INTERNET WAS DOOMED

The power of information would never be unleashed on the world. That was the consensus. The truth? No online database could replace your daily newspaper.

In the mid 90's, receiving an email that took two days to arrive, finally someone got to hear the American Online service provider "you've got mail" jingle. People didn't believe this mode of communication would ever scale to be effective in any business. It would take decades to build the infrastructure.

To better understand Bitcoin and Blockchain, there is a need to first understand some of the ground floor of where technology was, where it is today, some of the impact, and how it will change tomorrow.

Understanding Bitcoin's place in the history of the internet is useful if we want to form an understanding of where crypto assets are going, the Blockchain potential and what their future value might be.

Bitcoin is best understood as the first successful form of digital cash and digital hard money. It has the ability to perform the final settlement of large sums across the world in a matter of minutes. It has the characteristics of money, like unit of account, medium of exchange, and store of value. It doesn't rely on banks. It doesn't rely on governments. It relies on technology.

Blockchains let people send assets directly and safely to each other without going through a bank, credit card company, PayPal, or a big social media company. It is multiple databases in blocks that provide a platform for secure transactions, allowing everyone to know what is true in that transaction. Blockchain is open-source code. Anyone can download it for free, run it, and use it to develop endless possibilities of new applications.

This combination represents the most important technological invention in computer science of the last twenty years. How might this work out better than anything anyone expected, imagined, or planned?

We Need to Start with the Greatest Marvel in Technology to Date—The Internet

When I ask people what the difference is between the internet and the web, the most common answer is that they are both the same and really just bunch of computers linked together. Well, that is not true.

The internet is a network of networks that operates around the world, as if it were one. A huge global network of millions of interconnected networks—public, private, government or business networks—all speaking the same computer language or protocol. Rather, it is a suite of protocols which enable networked computers to perform a variety of end-to-end processes, or interconnected computers with an infrastructure of hundreds of thousands of servers (i.e. Internet Service Providers).

The WWW Web is just one service that rides on top of that infrastructure, where you search, get, and share information from web pages and websites. Other services that ride the internet are email, mobile—apps, file sharing, internet telephony and online games. So, the new technology in town that will also ride the internet is Bitcoin-the internet of money, with the underlying technology: Blockchain.When I ask how does the internet actually works, most know it is data servers connected around the world that allow a bunch of computers to connect.

But no one besides techies are going to drill down on the seven layers of data infrastructure, packet switching, the detailed technical standards for inter-computer communication, and how information can travel around the globe at 125,000 miles mph, or about 60% of light speed.

But, really, how does it matter today? It has become ubiquitous. Why would you need to know? The internet and web are full-on global mainstream for more than four billion people. So, I understand the question 'Why do we need to know?' It just works!

In fact, it works so well, it is entrenched in everything we do, impacting every business. I believe in ten years from now, Bitcoin, other cryptocurrencies, tokens, and Blockchain will be well passed the early adopter stage and into mainstream use. Just like the internet, it will work!

It was in 1989 that a British computer scientist, Tim Berners-Lee, submitted a proposal for a "distributed information system" in Switzerland. In 1990, the world's first website and server went live.

It was in 2009 that Satoshi Nakamoto submitted a paper for an open sourced project run as a peer-to-peer network called Bitcoin.

It is easy to look back at what the Tim Berners-lee accomplishment has brought to the world and say in hindsight, "Yes, of course, it makes sense". What we now can say is that what Satoshi has unleashed as an opensource network allows for whatever the imagination can build. What will it look like in twenty years?

Android & iPhone operating systems are open source which has allowed millions of new applications to be built. The internet was to be all-inclusive, allowing people around the globe to be connected on the information highway. It didn't happen.

We think the economy works best when it works for everyone. The internet was huge with its reach around the globe for all those who had access to the network. Despite its promise, the internet hasn't really delivered prosperity to all. Developing the costly infrastructure to developing countries is not something government or internet service providers were willing to do, leaving three billion people without internet or banking services.

Infrastructure

Many have no electricity. In sub-Saharan Africa, some 600 million people (almost two-thirds of the region's population) do not have regular electricity, and this applies to nearly a quarter of people living in South Asia.

Affordability

The cost of devices and connectivity is another factor preventing many people from benefiting. Broadband is only affordable for 100% of the population in just 29 countries.

Skills, Cultural Acceptance, Local Adoption, and Use

The vast majority (80%) of online content is only available in ten languages, which only about three billion people speak as their first language.

What if we had a world where the three billion others could live a step above poverty? That all could have an opportunity to create small business in ecommerce, creating new applications for all.

The reality is that existing poverty conditions that exist are so bad, war and government corruption will continue to prevent this from happening. Those differences indicate an important point about the prevalence of corruption, which is that there appears to be a close link between the level of fraud and bribery in a country and the political system it has. Corruption is much more likely to flourish where democratic foundations are weak and accountability is low. The new technology will be a slow process and not solve all of these problems, but what can happen when the many have control of their currency amongst their people is that it may take a hundred million out of poverty to have identity, access to financial services, and new opportunity.

The Internet Belongs to Everyone and No one

In 1994, there were thirty thousand websites globally. The internet bandwidth was incredibly slow, wasn't scaling fast enough, harboured illegal activity, and had a bleak outlook.

In 2019, Bitcoin is still viewed as a speculative investment, a bubble, a technology that allows bad guys to remain anonymous, has no regulation, and doesn't scale.

Understanding Some of the Basics

The internet will never be able to do video. Email will congest the bandwidth and bring the internet to a crawl. The web will destroy the internet. In 1992, you could spend a day on the internet and have explored all the websites. In 1992, a Netflix video of today would have stalled and broken the internet. There is a great site called Bitcoinobituaries.com where you can read the pronouncements of the death of Bitcoin since 2009.

Unlike the telephone network, which for years in most countries was run by a single company, the global internet consists of tens of thousands of interconnected networks run by service providers, individual companies, universities, governments, and others. Open standards enable this network of networks to communicate. This makes it possible for anyone to create content, offer services, and sell products, without requiring permission from a central authority.

The internet is a network of networks that needs to operate around the world as if it were one. Domain names give global coordination of the Internet Protocol, addressing systems as well as routing all Internet traffic. This is Transmission Control Protocol," or TCP. (The acronym we use to refer to these today is TCP/IP.) The protocol known as "the 'handshake' introduces distant and different computers to each other in a virtual space.

Has the Internet Changed or Does it Need Upgrades?

Currently there are two types of Internet Protocol (IP) addresses in active use: IP version 4 (IPv4) and IP version 6 (IPv6). IPv4 addresses are thirty-two-bit numbers and IPv6 addresses are 128-bit numbers. The major change here is that there were concerns about running out of internet addresses with Version 4, and after two decades, the migration to IP6 is still in play. Two of the biggest problems with the internet protocol was that it did not build in identity and it did not build in encryption. Technology developers wanted and tried to build this in but back then consumers did not care about security at that time. The Internet has trouble with preserving our privacy, establishing our identities with issues on security, and we still don't know if we're dealing with a bot or a person.

The internet needs a profound shift in how it can be used to create new forms of value. Security issues are high, and data breaches are happening every day. Companies know that it is not a matter of 'if you get breached but when'. The Blockchain will bring trust—a trusted relationship, without ever meeting each other or doing business together.

New protocols have been developed such as the Lightning Network. It is like another application layer which operates on top of the Bitcoin network as a cache to allow payments to flow without congestion by not immediately putting them on the Blockchain. The Lightning Network scales blockchains and enables trustless instant payments by leveraging the security of the underlying blockchain. It is an open protocol layer that leverages the power of blockchains and smart contracts to make cheap, fast, private transactions available to anyone around the world.

Like Bitcoin itself, the Lightning Network is decentralized:

- No online payment center or credit card company is required to authorize transactions or to hold funds. Instead, users directly connect to the Lightning Network by creating multiple payment channels of their choice, without relying on a centralized third party.

- The Lightning Network is a protocol that aims to improve Bitcoin's scalability and speed enabling ultra-fast and nearly costless transactions atop the Bitcoin Blockchain.
- In January, 2018, Blockstream launched a payment processing system for web retailers called "Lightning Charge". These additions make it easy for developers to build apps on the Lightning Network and to use c-lightning to create their own, independent web-payment infrastructures.

As with any new technology, there have been dangerous bugs in Bitcoin's Lightning Network which have since been fixed. There will always be bugs. What matters the most is how to handle this in the best way to protect users.

While the internet has been spread into our society, thousands of new cultures and innovations have been developed by people utilizing freedom space such as e-mail, the web, and social media. The internet was to bring inclusion for all to access the world of information held only by institutions. However, we have had several rights violations limitations on freedom, and abuses of control.

To what extent will the Internet be used as a tool to influence and control society, rather than promote freedom and human rights? Will the risks of the internet to freedom and rights outweigh the benefits to society and the individual? If so, what reaction will this evoke?

Let the internet do money instead of government.

Two Major Components of Blockchain

- Algorithms—The method of solving problems with mathematics and science.
- Cryptography—The evolution of secret messages.

In its basic form, encryption changing content to something that is unrecognizable to those who are not authorized to consume it. The recipient of the message has the codes to decipher it.

There has always been a desire by civilization to hide secret messages; then, one hundred years ago, a systematic study of cryptology as a science began.

Bitcoin and Blockchain use cryptography and algorithms to secure the network.

The Historic Beginnings

Julius Caesar was known to use a form of encryption to convey secret messages to his army generals posted in the front lines. The Caesar cipher was a method used to make a substitution for each character of the plain text by another character, in order to disguise or encrypt the actual message. To reveal the actual message, these ciphers depended on the secrecy of the system. Once the system is known, these encrypted messages can easily be infiltrated and the system exposed.

In the Second World War, The Enigma machine was invented by the Germans and heavily used by them. It was like a typewriter with rotors rotating at different rates as you typed. The words would show up as a form of gibberish unless the message was delivered to the recipient with a machine of their own. The Enigma machine's cipher was eventually broken by Polish cryptographers. This, in turn, led to the interception of daily strategic German updates, creating many significant turning points for the allied forces in the war.

In the next forty years, cryptography attracted commercial attention, with businesses trying to secure their data from competitors.

Who knew that an algorithm was part of everyday life?

The Benefits of Algorithmic Thinking

Algorithmic thinking, or the ability to define clear steps to solve a problem, is crucial in subjects like math and science. The word "algorithm" is all around us. Algorithms are everywhere, governing everything from the technology we use to the mundane decisions made every day. Algorithms are fascinating and, although some are quite complex, the concept itself is actually quite simple.

What Is an Algorithm and how does it apply to Bitcoin and Blockchain?

An algorithm is a detailed step-by-step instruction set or formula for solving a problem or completing a task. In computing, programmers write algorithms that instruct the computer how to perform a task. Your last recipe you followed for food is an algorithm, the method you use to solve addition or long division problems is an algorithm, and the process of folding a shirt or a pair of pants is an algorithm. Even your morning routine could be considered an algorithm! Kids can write their own algorithms! The school system which does not teach anything about money, finances, or task-orientated processes should talk about algorithms. You can have your child write out their morning algorithm, like brushing their teeth or eating cereal. Without knowing it, they'll actually be exploring important computational concepts like repetition (brush bottom left teeth five times), sequencing (put cereal in bowl, and then put in milk), and conditional logic (if the bowl is empty, stop eating).

It is a process towards understanding the mathematics of how things work.

Modern cryptography is heavily based on mathematical theory and computer science practice; cryptographic algorithms are designed around computational hardness assumptions, making them hard to break by any adversary. Encryption keys are needed to encrypt and decrypt messages or data. Cryptography is the math of keeping secrets. Cryptocurrencies are digital currencies that rely on cryptographic technology to regulate each generation of the currency and verification of respective transactions.

In the nine years that Bitcoin has been in existence, it has never been hacked, and that is what makes it an incredible store of wealth. Like gold and silver, cryptocurrency derives its value from supply and demand. Value is not controlled by any country, central bank, or other single authority. So, Bitcoin and cryptocurrencies have no physical form; the network generating the cryptocurrency is completely decentralized, with all transactions performed only by the system users. Transactions are irreversible and generally pseudonymous—that is, the transactions are linked to a consumer's public key, not to an individual. The transactions use cryptography for security and anti-counterfeiting measures.

Cryptocurrencies are based on the concept of money as information: the money is a string of bits, sent as a message. Most people have seen the term 256-bit encryption. This means the data is scrambled and makes it unreadable. It is one of the most secure methods we have today. When you encrypt something, you're taking the 'plaintext', and performing an algorithmic function on it to create a piece of encrypted 'ciphertext'. The algorithm you're using is called the key. Only another party with knowledge of the associated private key can decrypt it with encryption keys.

These keys are private and public encryption keys. Using a private key, the same key (secret key) is used for encryption and decryption. This key is symmetric because the only key is copied or shared by another party to decrypt the cipher text. It is faster than the public key cryptography. With a public key, two keys are used; one key is used for encryption, and another key is used for decryption. One key (public key) is used to encrypt the plain text to convert it into cipher text, and another key (private key) is used by the receiver to decrypt the cipher text to read the message.

So, let's look at those 256-bit keys. A 256-bit key can have 2256 possible combinations. A two-bit key for example, would have four possible combinations and be easy to crack.

We're dealing in exponentiation here, though, so each time you raise the exponent, you increase the number of possible combinations wildly. 2256 is 2 x 2, x 2, x 2, 256 times. A 256-bit private key will have millions of combinations. The best way to crack an encryption key is 'brute-forcing,' which consists of an attacker using many computers and/or software applications, directing thousands of attempts with passwords or passphrases with the hope of eventually guessing correctly—basically trial and error. A hacker must try most of the 2256 possible combinations before arriving at the conclusion, and the time it would take to do this would last way beyond any human lifespan.

Long story short, the biggest threat to your encryption and your encryption keys is still mismanagement; the technology behind them is sound. No Super Computer on the face of this earth can crack the code—period. It makes it computationally infeasible for data to be stolen.

3

WHY YOU MAY NOT KNOW MUCH ABOUT MONEY

In a four thousand-year-old snapshot of value and money, we can gain insight into a tradition of debt cancellation in Mesopotamia and Egypt from 3000 to 1000 B.C. The kings at this time were able to record tax payments form the peasants. This was drawn out on stone tablets. Hammurabi, king of Babylon, had one such set of debt cancellation records. Was this the original form of Blockchain as we know it today?

The characteristics of money are the same today as they were four thousand years ago. There arose a need for people to exchange the surplus goods they produced for those they did not have. These financial transactions, or promises, edged on clay did not have inherent value but they functioned as money because the parties involved in the transaction trusted it.

The ledger recorded the taxes paid or unpaid, and when debts could not be paid the king would seize the peasants' property. If peasants were unable to pay off their debts, they could also find themselves reduced to the condition of serfs or slaves; indebtedness could also lead to members of their family being made slaves. But in order to ensure social peace and stability, and especially to prevent peasants' living conditions from deteriorating, the authorities periodically cancelled all debt, effectively 'wiping the slate clean'.

Such proclamations of debt cancellation were an occasion for great festivities, and the tradition of destroying the tablets upon which the debts were inscribed was inaugurated. The public authorities kept a strict record of debts on tablets conserved in the Temple. So, money has evolved to be what it is because people trust it.

The first thing to know about Blockchain: it is a distributed ledger. The term 'ledger' came from the church. It was when they kept an account of the number of bibles that were in a church. It was very important to keep an account of how many they had, and they stored them on a ledger for easy access.

The ledger is an account of things you want to keep track of. There are accounting ledgers, product ledgers, and so on. Because of the way technology has evolved, most companies keep a copy of their own data by themselves and they don't share.

The Money-counting Problem of Europe

In 1200 A.D., the use of coins as money in Europe and Asia were introduced, and independent states engaged in commerce with each other.

However, each state developed its own coinage, creating trading issues where trades would use as many as seven different coins.

This really complicated business, as even the smallest transactions required complex calculations. The Roman numerical system couldn't do complex mathematical and financial calculations, and there was a need for merchants to keep records of the money, leading to the invention of the abacus, a counting machine.

Medici Family & The Beginning of the Central Banks

The dictionary definition of credit (n.): 1540s, "belief, faith," from Middle French crédit (15c.) " belief, trust," from Italian credito, from Latin creditum "a loan, thing entrusted to another," neuter past participle of credere "to trust, entrust, believe".

In Tuscany, Italy, at the time of the Renaissance Era in the years 1300 to 1500, there was a revival from the Dark Ages into art, cultural, and political reform. This was a transformation created by the most powerful family at that time: The Medici Family.

The Medici's were technology disrupters of their time, who figured out a way how to be intermediaries between savers, borrowers, and traders. They created a high-powered system of money-creation, money not being a physical currency but a system for organizing, expanding, and sharing society's debts and payments. It was built on trust.

By creating this centralized system of trust and putting themselves in the middle of it for a fee, banks became extremely powerful, since strangers could not do business without the banks. This management of currency and asset exchange dates back to when banks first assumed dominance in the monetary economy of Europe, and now some six hundred years later, we now know that banks and the financial institutions hold the power.

The Latin word 'Credo' means 'I believe', evolved to be 'credit'. It was first used by money-lenders when advancing money to merchants. By the early 16th century, as money became increasingly accepted as a measure and storage of value, almost every kingdom and state across Europe and Asia began producing its own money. This allowed more credit, enabling more merchants to take bigger risks and travel more widely to trade. It made way for an explosion in trade which, in turn, created the wealth and capital that would finance the projects from which great civilizations would grow and conquer the world.

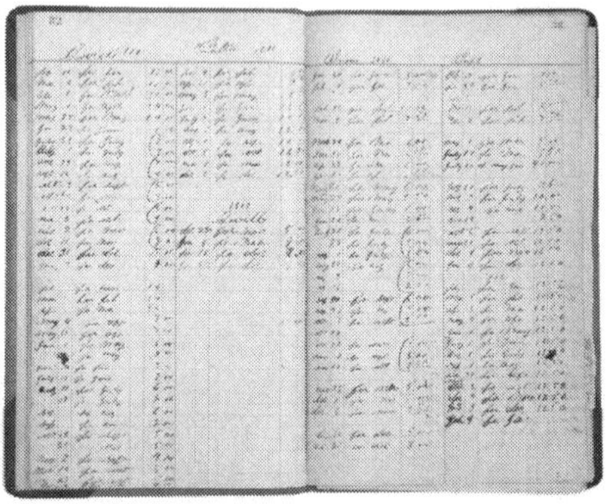

Gradually, the technology of bookkeeping evolved, and within no time, banks were using the double-entry system to maintain their checks and balances, keeping their own detailed ledger. With double-entry bookkeeping, every accounting transaction has two effects on a company's finances. The general ledger is the record of the two sides of each transaction.

There's a bit of a problem though in the world of digital ledgers: a digital coin or token of money can't be spent twice.

Because what would happen if more digital coins were created and added to one's balance whenever they wanted? What would prevent one from sending the digital coins to several people? How can one just hand over my digital coins to someone in the usual way?

The Solution

Satoshi Nakamoto created this combination of a number of different technologies that allowed digital currency to be issued and exchanged without needing a central authority or a clearing house, solving the problem of double-spending. You can email or send the currency anywhere you want and transmit on any network, but every participant on the network can check that the money had not been spent before and could trust that transaction without a central authority. What if we gave this ledger to everybody? Instead of the ledger living on a central computer server, it'll live in everybody's computers. All the transactions that have ever happened, from all time, will be recorded in it.

You can't cheat it. You can't send you digital coins you don't have, because then it wouldn't sync up with everybody else in the system. It'd be a tough system to beat. It's called the Bitcoin protocol. And those digital coins are the Bitcoins within the system.

What would a public ledger enable?

- It's open source
- When you make an exchange, you now know that digital coin certifiably left my possession and it is now completely yours. It will be updated and verified by the public ledger.
- Because it's a public ledger, you don't need a third-party to make sure you didn't cheat, or make extra copies for others, or send twice, or thrice.

Within the system, the exchange of a digital coin is now just like the exchange of a physical one. You can send it with the click of a button, and you can still drop it in your digital pocket if you were on the other side of the world. You can even make other digital things ride on top of these digital coins. Maybe attach some text on it—a digital note. Or maybe attach more important things; like, say, a contract, or a stock certificate, or an ID card. This capability is going to create an explosion of new developments.

John Maynard Keynes said, "The banking system and how money works is purposely complex. By this means government can secretly and unobserved confiscate the wealth of the people and not one man in a million will detect the theft." You might first learn how the world monetary and economic system works. Good luck with that, though. It would be like taking a full-year university course at a minimum. Cash is just a piece of paper backed by no one.

Money represents an abstraction of value; it's a way of communicating value. It's a language. Bitcoin is a language, a form of communication, a store of value, and a new form of money.

We use it as the basis of social interaction because by communicating value to each other, we create social bonds. So, money is also a very important social construct. This is an ancient technology. Yet, ironically, it's one of the technologies that is least-studied from a historical and technologic perspective.

There have been a lot of concerns about the first generation of the internet. One of the biggest fears is that powerful corporations will capture most of the technology and use it to fortify their private empires. This has already proven true with much of our digital experience. We use proprietary stores to acquire and use new apps on our phones, tablets, and now watches. Search engines target us with countless ads. Consumers are fully transparent.

Banks are naturally among the oldest secret-keepers. They make good judgments about whom to lend to and how to process payments, because they have access to private information.

They get that information by promising to keep it secret, giving them an advantage that other industries don't possess.

Blockchain technology has emerged as an important global resource, but it's also a disruptor. This quote from an English historian rings true: "Power tends to corrupt, and absolute power corrupts absolutely."

Bitcoin and Blockchain stops this. We look at Bitcoin today and it represents an invention, a new form of money. One of the most popular forms of these abstractions was to use precious metals to express value. Precious metals combine some of the most important characteristics of money: hard to find (scarce); easily transportable (at least when compared to a giant rock or a whole barrel of feathers); easy to divide (you can cut a gold coin into pieces and subdivide the pieces); and universally valued for aesthetic purposes.

That's the second major transformation in money technology. It took hundreds of thousands of years before we saw the introduction of precious metals.

This new invention of money as paper was very controversial. The same is true of cryptocurrencies.

The dramatic change in the past was when instead of trading in gold, they would trade in pieces of paper. For a lot of people, this was unthinkable. How could a piece of paper have value or hold value? It was flimsy, could float away or easily get torn or lost. It took about four hundred years for paper, as money, to become accepted broadly. It was a big aberration.

Paper to Plastic

Then, about sixty years ago, we saw a new form of money in the form of plastic cards. In fact, the first cards were paper again. In the United States, Diners Club was the first to create a credit card. Then, people took that and they said, "This isn't money. Why don't you give me some of the good old paper money that I know?" That was another big transformation in money.

Plastic to Bitcoin

So, what is Bitcoin? The fundamental issue in describing Bitcoin is that if you use references to our existing experience, that experience is based on thousands of years of understanding what money is in a very physical form. Now, we're trying to explain a form of money that is completely abstract.

Most people think that it's simply a payment system, and that Bitcoin is simply a form of digitization of money. It's digital money, right?

Great. Kind of pointless because we already have digital money. We all use digital money every day, and had been long before Bitcoin came into existence. You have bank accounts. Those bank accounts have digital ledgers. We use those bank accounts to send payments electronically. Trillions of currency are moved around each day. That's digital money. Money around the world is created when commercial banks extend a loan. It is created out of thin air which should have no financial value whatsoever. None of the money we use today is backed by anything other than a promise by a bank or government to honor the value of a paper or coin. And if we wanted to convert to cash we can't because only 3% is available.

Bitcoin is a fundamental transformation of the technology of money.

Bitcoin isn't just digital money. It's difficult to grasp because it's so different from everything we've known before. Bitcoin is happening at a moment in history when we're seeing a transformation of many fundamental social institutions. Telecommunication networks with their own infrastructure now deliver their platform on the web. So, we moved from institutions to platforms. Then, we started seeing an even more important transformation, when we moved from platforms to protocols.

The interesting thing about the change between a platform and a protocol is this: when you have a protocol there is no central appeal. TCP/IP doesn't work in reference to a service provider. TCP/IP works without context everywhere in the world. You don't have to sign up for an account to use TCP/IP; you just have to use the language. Once you move from a platform to a language, it opens up all of these possibilities.

Bitcoin is the first network-centric, protocol-based form of money. That means it exists without reference to an institutional or platform context.

Peer-to-Peer Architecture

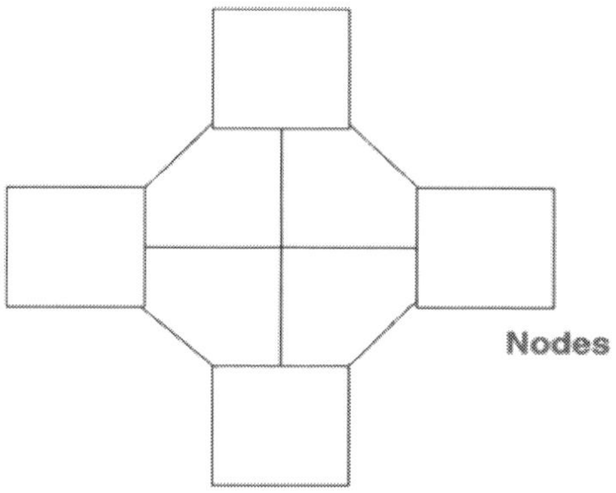

Peer-to-Peer Network Model

We say that Bitcoin is peer-to-peer money. What does that mean? It refers to an architecture used in terms of computer science, or networking, or distributed systems, to describe the relationship between participants and a system. The architecture of Bitcoin is peer-to-peer because every participant in the network speaks the Bitcoin protocol on an equal level. There are no special Bitcoin nodes; all nodes are the same.

Peer-to-peer means that when you send out a transaction to the network, every peer treats it the same. It has no context inside the peer's system other than what it gets from the network. An interesting issue in distributed systems is this issue of context and state.

If you log in to Facebook, you're not using a protocol. All of the 'state' is controlled by Facebook. You have a login session and all of the data is held by them. This is called client-server architecture. Bitcoin is different because it's peer-to-peer, just like email or TCP/IP.

Client-Server Architecture

This is similar to a bank account You are already a bank, because your deposits are loaned to the bank and they turn around and lend to others via mortgages, commercial loans or credit cards.

This is a client-server relationship, because that money only exists as a form of debt in a ledger that you do not control. it's a ledger that is stored by a server, and you are simply a client. In fact, you have no control over it at all.

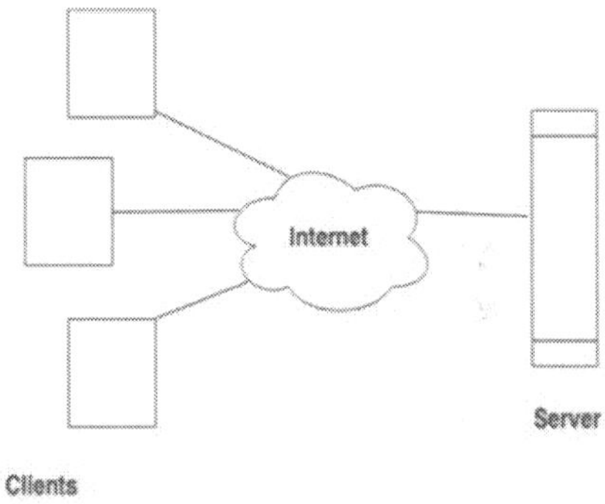

Client-Server Network Model

Blockchain

Architecture: A blockchain is a network of computers connected not to a local server but rather to one another. Transactions are put together and then sent to each node in the form of block. Since new transactions are distributed across the network, they are independently classified and processed by every node.

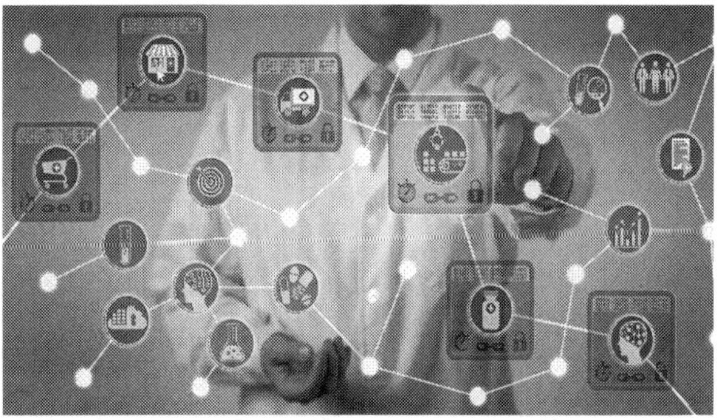

There is New Money in Town!

It has only taken fifty years for credit cards to move from paper imprints to phone authorization, swiping and chips to tapping phones. That is considered advanced technology by the banks.

After the major breakthrough and adoption of credit cards chip cards were originally conceived by EMV (Europay, MasterCard, and Visa). They are more secure than traditional debit and credit cards, because account information stored on cards is encrypted uniquely each time it is accessed.

Chip credit cards can still be "hacked," in a sense. If a thief inserts a "skimming" device into a credit card terminal, they can copy data from your credit card and later make a copy of the card. However, skimmers can only copy data from your card's magnetic stripe, not its chip, which is much more encrypted.

Yet, despite the security that chips provide, all chip cards still come with magnetic stripes. They're necessary because only 70% of merchants have switched over to payment terminals that can process chips. If a merchant does not have a chip card reader, you'll have to swipe the card. This deprives you of the extra security protections built into the chip and leads to a greater risk of fraud.

Bitcoin debit cards are secured using EMV contact or contactless technology. Three prime examples are BitPlastic, Coinkite, and Cryptopay. They are marketed as mechanisms for anonymously withdrawing funds and shopping. To be interoperable with the card networks, banks issue the debit cards branded as Visa or Mastercard cards.

SHIFT CARD is one example.

Coinbase is a Bitcoin exchange company. Coinbase allows consumers to store Bitcoins; consumers can use Coinbase to exchange Bitcoins into fiat currencies (cash).

Very few merchants accept cryptocurrency as payment, making it difficult for consumers to make regular purchases using cryptocurrency. The Shift Card is a Visa debit card that currently allows a cryptocurrency Exchange like Coinbase to provide users in 24 U.S. states and territories to spend Bitcoins anywhere Visa is accepted. A payment transaction using the Shift Card works as follows:

- A consumer swipes a Shift Card at a merchant POS.
- The equivalent value in Bitcoins is debited from the consumer's Coinbase Bitcoin wallet. The Bitcoin value at the time of purchase is based on the current spot price of Bitcoins on Coinbase. There is no transaction fee for converting Bitcoins to fiat currency.
- When the payment transaction is approved, the merchant receives payment in U.S. dollars.

Shift cardholders can therefore spend Bitcoins at over 38 million merchants worldwide. Naturally, users would most likely spend their Bitcoin on one of it's high volatile rides.

Old Money Closes at 5PM

You wouldn't believe what the future holds. What more is possible and how did we get to be so lucky?

The new money is not a dirty word. New money isn't paper–it's progress. Old money closes at 5pm. New money is always open every day. Old money is stuck in the past. New money gets better every day. New money is for all people. You can't stop things like Bitcoin. It will be everywhere and the world will have to readjust. These are times of great challenge, great excitement for inclusion for the unbanked and one of great opportunity for new innovations to come.

But when will cash disappear? Sounds like it is pretty soon. First World governments around the world want it sooner than later. They want it gone so they can track and tax anyone and everyone. Developing countries are the opposite, due to prevalent and inherent corruption.

Enter the new money in town! When Bitcoin evolves past the early-adopter stage and matures, it won't need POS systems. You can send money to anyone around the world instantly. Your device becomes your own bank

The Federal Reserve, or the Central Bank of the US, was created to provide stability to the financial system. The Four Horsemen of Banking (Bank of America, JP Morgan Chase, Citigroup, and Wells Fargo) own the Four Horsemen of Oil (Exxon Mobil, Royal Dutch/

Shell, BP, and Chevron Texaco), in tandem with Deutsche Bank, BNP, Barclays, and other European old-money behemoths. Change and adoption is already happening.

The 2008 crisis with MBS (mortgage backed securities) in the US created a financial crisis. For almost a decade, the government and financial institutions offered new homeowners easy access to money to buy homes before the mortgage crisis emerged. Borrowers got into high-risk mortgages. Easy money for those that before could not afford homes. Prices started falling in 2006. A long chain of events happened. Borrowers who bought more expensive homes than they could afford eventually stopped making mortgage payments. The economic environment changed dramatically, and eventually the Federal Reserve bought 1.75 trillion dollars worth of real estate with currency they whipped up. They instantly became the largest real estate holders in the world.

4

BITCOIN – THE AGE OF NEW MONEY

1397Yc3SW6TKmk2cKcxLgkKWoBzka7jPoZ (Bitcoin)

Think of this like a bank routing number, complete with your bank account number where people can wire you currency. Using it, you can carry out transactions using digital currency. Bitcoin is a transaction that has value attached to it. Bitcoin has no physical form. A special string of alphanumeric characters represents a Bitcoin public address.

1397Yc3SW6TKmk2cKcxLgkKWoBzka7jPoZ

With Bitcoin, most of the terms are very new and confusing, or in a sense misleading, which further adds to the confusion. Terms like a 'wallet', where cryptocurrency is stored or lives, 'mining', where people worldwide are running software that confirms new blocks of recent transactions into the permanent blockchain, and 'Hodl', meaning you hold Bitcoin for long-term strategy.

So, it's not just a currency. It's a trust application at work. Currency is a tiny part of it. It is a transformation of money and difficult to grasp. It's a currency in itself but is also the name of the network that supports the currency. It is a store of value. It is also a set of protocols and standards. The decentralized control of each cryptocurrency works through distributed ledger technology, typically a Blockchain, that serves as a public financial transaction database.

A digital asset, in essence, is anything that exists in a binary format and comes with the right to use. Data that do not possess that right are not considered assets. Digital assets include but are not exclusive to: digital documents, audible content, motion picture, and other relevant digital data.

We can look back at the Pokémon story to understand what the store of value meant for hundreds of thousands of children. The Pokémon phenomenon started very simply, as the hobby of a man named Satoshi Tajiri, who enjoyed catching insects and tadpoles near his home as a child. When Tajiri discovered the Game Boy and Game Boy Game Link Cable, he thought of insects traveling along the wire, moving with the cable. Tajiri also took a concept from a main character who used giant monsters held within small capsules to help him fight. This later evolved into Pokémon. It was these two concepts that gave him the idea for a game called 'Capsule Monsters', which was the first iteration of what we now know as Pokémon.

The Pokémon trading card game has shipped more than 21.5 billion cards to seventy-four countries in ten languages. Pokémon was relatively harmless, as long as children didn't start to confuse the world of the game with the real world in which they lived the rest of their lives. The more Pokémon you have, the greater power you possess (the slogan is GOTTA CATCH 'EM ALL). And never underestimate a child's ability to master the Poké arcana required to accumulate such power: the ease with which they slip into cunning and thuggery can stun a mergers-and-acquisitions lawyer.

What it really had was a store of value. It was a currency on its own. Kids believed there was so much value in having certain cards, creating a strong desire to acquire them through bartering, trading strategies, and the acquisition of certain cards. Kids played the game, trading cards feverishly, trying to acquire cards that had special powers.

These had so much value to them and they would badger their parents to buy them more packs of cards, fixated on cards rather than schoolwork, go into trading mode with school mates, trade their lunch, a coat, a bike, or whatever they deemed to be of less value. Even cash was of less value to them, besides the fact that it would allow them to purchase more cards. In the minds of kids around the world, Pokémon was a currency.

Around 97% of "money" around the world is created when commercial banks extend a loan.

All of this further confirms that most money is created out of thin air which should have no financial value whatsoever. None of the money we use today is backed by anything other than a promise by a bank or government to honor the value of a paper or coin. And if we wanted to convert to cash, we couldn't, because only 3% is available.

Well, if everyone were to ask the powers that be to make good on that promise simultaneously, the entire economy would collapse in a heartbeat. Ever since the gold standard was dismissed to give money its actual value, things have unraveled pretty quickly.

Bitcoin is the most intriguing new financial instrument of the decade.

It was an early implementation of a technology called a Blockchain, a database network that records financial transactions independent of any central bank, and it could replace bankers and brokers within a decade.

- Gold is often classified as a commodity; however, it behaves more like a currency and is not tied to any country.

- Fiat money (cash) is a currency that a government has declared to be legal tender tied to that country.
- Bitcoin has no physical form. It uses a new value model with protocols and standards which currency is only a part of. It is not tied to any country.

So how is Bitcoin created?

Bitcoin is based on Proof of Work (PoW) algorithm, the computer language used to make complex calculations. Bitcoin is able to allow a completely decentralized network of computers to agree on what transactions have occurred on a network. The people or miners act as auditors verifying transactions, essentially agreeing on who currently has the money. By verifying transactions, miners are helping to prevent the "double-spending" problem, or in effect not using the same digital currency twice.

The first miner or group of miners to arrive at the right answer to an algorithmic problem is rewarded with a quantity of bitcoin. This process is also known as proof of work. This is known as a "block" that becomes part of the Blockchain. The Blockchain is basically the database, the system of transaction bookkeeping where all cryptocurrency activities are recorded.

The purpose of the reward is an incentive which motivates people to assist in the primary purpose of mining: to support, legitimize and monitor the Bitcoin network and its blockchain. These responsibilities are spread among many users or nodes all over the world.

Computers solving these calculations are awarded tokens as a result of their efforts.

Bitcoin is the Godfather of cryptocurrencies. Other currencies and tokens create value on their own as they focus on a specific space, industry or application.

Many have features which Bitcoin does not have and may be more applicable to a certain industry. Each cryptocurrency is traded in its own general ledger or Blockchain. They have names such as Ethereum, Dash, Monero, LiteCoin, DodgCoin, and Tron. Ripple, for example, was designed for the international trading of goods and services. Ethereum will empower the financial institutions. Each has its own Blockchain software.

Key Bitcoin Attributes

- Not centrally controlled by any one person or group
- Alignment of incentives
- Simultaneous protocol, network, and currency
- Voluntary participation
- Users are 100% responsible
- Distributed and decentralized
- Economically infeasible, hard to break algorithms

We have a whole world economy which is built on centralization.

A Bitcoin's value is derived from a continuous auction process with every Bitcoin trading house. As a result, a Bitcoin's value can vary from place to place. Where can you get the most money for your Bitcoin? For example, because Zimbabwe's native currency is so distrusted and because many of Zimbabwe's citizens now have Bitcoin wallets on their phones, there is high demand for the cryptocurrency in the country and very limited supply.

The Blockchain is the underlying technology that makes Bitcoin possible. Money nowadays is numbers and spreadsheets. You can't transmit currencies without a central bank. That is what a Blockchain is. The way to record the fundamental data of transactions and commerce, just like what money does and what accounting does through decentralized networks.

The major factor is we have not been able to do this securely on decentralized digital networks or the internet. And because we never had this before, we had to have centralized ledgers that were curated and organized and managed by governments, by large corporations and centralized authorities that we needed to trust.

This is where this notion of a 'trustless' ledger came from—bypassing the need to pass transactions and information through a centralized system or node that we needed to trust. This is what this was originally invented for. This was a new way to put money on the internet in such a way that it didn't require central banks, and it didn't require governments to issue it.

A more fundamental problem is that it enabled a distributed network of a computers to agree on a state of reality on transactions and who has the money without having to rely on trust. It created the first ever de-centralized trust network. The first application is currency, but it also solves anything that requires trust you can now do without a central authority.

Internet—Decentralized

The internet was the first major example of the use of technology to solve problems such as information, facts, truth and opinion that were centralized and controlled to an open global forum in a decentralised way that gave access to everyone.

This allowed individuals to exchange information communication and truth without having to rely on a central party, most importantly removing control from centralized institutions that had control of information, the origin and definition of truth on a cultural basis.

The internet allowed people to drive their own truth without an authority. The internet was centralized to some degree at the beginning but expanded and scaled as more participants came into the fold. There is a network effect, like Facebook for example. It multiplies the value of the network very rapidly.

What Makes Bitcoin So Unique

There are 150 other currencies that use the basis of the Bitcoin architecture. They have the opensource technology and have built different applications that solve specific industry problems. There are other types of cryptocurrencies or tokens numbering some 1500 or more, depending which ones have disappeared already and which ones are being created.

The total amount of Bitcoin is twenty-one million. The unit breakdown of Bitcoin is similar to a nickel or a penny on a dollar except the units are much smaller. The 'satoshi' is the smallest unit that is recorded on the bitcoin blockchains: One satoshi represents a decimal. There are 100,000,000 satoshi in a BTC. Each bitcoin (BTC) is divisible to the 8th decimal place, so each BTC can be split into 100,000,000 units. There are many converters that can be found online: *https://www.buybitcoinworldwide.com/.* It offers a store of value and is viewed as a scarce and finite resource, and thus an intriguing asset to consider purchasing as the dollar falls.

Is it inherently deflationary in the long-term? Bitcoin is the first trans-national currency. The government is allowed to print an infinite amount of money. With Bitcoin, twenty-one million is the limit.

Bitcoin is not just another currency in the world. It is not in a closed ecosystem like each currency each country has. If Venezuela could pick and convert any time into any currency they wanted, there would be a different monetary problem.

Gold has been used as an investment and is often classified as a commodity; however, it behaves more like a currency. Gold is a global monetary asset, and its price reflects the global sentiment.

Fiat money (cash) is a currency that a government has declared to be legal tender, but is not backed by a physical commodity. The term derives from the Latin fiat ("it shall be" or "let it be done"), as fiat money did not spontaneously emerge in the free market, but was established by government regulation or law.

Ethereum

Ethereum cryptocurrency has the largest market share after Bitcoin. Financial institutions are most interested in what it solves and are investing billions of dollars.

It was built on the foundation of Bitcoin and enhances the potential to move value without a central authority. Realizing that Bitcoin has an issue scaling to handle large volumes, specific-use cases for business, financial institutions and government would need their own central authority to control them but also be built with Blockchain. The significance here was it was identified that in order to build decentralized applications with the Bitcoin Blockchain, many changes to the programming code would need to be made—even a new Blockchain for high-volume transactions.

In 2014, they launched a crowd-funding sale of its cryptocurrency token. The first release of the Ethereum network was a basic platform that only the more technical savvy could use to build their own applications. The application template allowed for any developer to build, and now, having gone through many phases, Ethereum has become intuitive and user-friendly for new applications to be applied.

The E protocol can now do almost anything that average programming languages can do, which means the number of applications that can be built into the E Blockchain ecosystem is enormous. Any company or government process could be rebuilt as a digital representation based upon Ethereum's groundwork. The applications are endless: proof of work, transparency with autonomous businesses, the ability to help parties come to agreement and work together for the common interest, commerce—you name it.The Enterprise Ethereum Alliance (EEA) was created with a goal of allowing enterprises the opportunity to take advantage of the scalability, security, and confidentiality of Ethereum's blockchain technology. Some of the founding members of the 150 member plus included **JPMorgan Chase, Microsoft, Cisco, Mastercard, Intel, Accenture, BP, and Credit Suisse.**

Imagine twenty years from now, a driverless car cruises around in search of passengers. After dropping someone off, the car uses its profits for a trip to a charging station. Other than its initial programming, the car doesn't need outside help to determine how to carry out its mission.

The passenger only pays for the ride which would incorporate all costs just for that piece of the journey. Insurance, license, gas, depreciation, all built into that one ride and paid with digital currency on the Ethereum Blockchain.

Ripple

Ripple is currently one of the top five cryptocurrencies. Ripple's aim to improve cross-border transfers through distributed ledger technology will have a significant effect on commerce and finance, and thus a positive impact on the world economy.

Improve Cross-border Transfers

Few dispute the fact that the current system of transferring money across currencies and borders is inefficient and wasteful. In order to be able to make cross-border payments on behalf of their clients, commercial banks need to have accounts with banks in the destination country. They generally hold balances of that country's currency in these accounts. Often the balance just sits there, waiting for a payment instruction. Part of Ripple's proprietary technology uses a distributed ledger—a platform used by banks and other financial institutions to settle cross-border payments. It uses a messaging protocol that extracts the information necessary to arrange the payment. Once the pre-transaction validation of the accounts and balances have been completed, all parties put a lock on the funds.

At this point, cryptographic signatures are generated that verify the funds are available and that the parties are ready to make the transaction. The end result is creating efficiencies, security, and saving millions of dollars. It has its own ecosystem where transactions are validated by a list of "trusted" nodes, authorized by Ripple. All one hundred billion tokens were issued to Ripple in 2015. The creators kept twenty billion and gave the rest to the company, which has been using it to develop liquidity, methodically distributing them to clients and incentivizing market making activity.

Ripple's platforms are being used or investigated by over one hundred financial institutions, along with a handful of payment providers, such as Western Union and Moneygram. Mexican money transfer firm Cuallix is actually using it in production.

So, what should we expect from Blockchain?

There's little doubt about the potential that this technology holds. Industries like finance and healthcare already see its impact. The changes will not happen in the next five years. It will take its own time to slowly but steadily seep into all our structures for good and to change them from within.

The Blockchain is here to stay. However, the world needs to start small to correctly understand and structure how and in what ways do we want to evolve Blockchain.

Blockchain shows a glimpse of the world that could be—the beginning of a system wherein transactions are recorded, stored, and verified on a digital network. Every record has a timestamp along with a signature, and no data can be tampered with, recreated, or in any way corrupted. A self-sustaining system to the point where intermediaries like lawyers, brokers, and bankers will not be needed anymore. Different parties will transact directly with one another with little or no friction whatsoever. That is the future that Blockchain promises.

Blockchain was missing an important piece of what Bitcoin now solves.

What is Bitcoin? CASH VALUE. The first point was understanding money. The second is understanding control of the banking industry.

So back to the question of what Bitcoin is: it's still being defined!

It is completely abstract. It is only a digital footprint on the internet. It is a technical language, a currency. Crypto is good money. Monies come and go throughout history, but Bitcoin possesses all the attributes of good money and remains in a league on its own. Government can print money anytime and put it into circulation. Bitcoin exists as a ledger entry backed up by thousands of computers around the world.

Bitcoin doesn't care if the Bitcoin address is the address of a multimillionaire, the address of a central bank, the address of a smart contract, the address of a device, the address of a human, or a bad actor. It doesn't know. It doesn't care if the transaction is carrying lots of money or not much money at all. It doesn't care if the address is in Thailand or downtown New York.

It moves money from one address to another based on a simple locking script. And that means that if you want to build a new application on top of Bitcoin, you can upgrade the devices and you can build an application.

You don't need to ask for anyone's permission to innovate. Write the app, launch it on your endpoint, and Bitcoin will route it.

Never Mind Bitcoin—What Do You Know About Money?

What do you know about how money works? We don't even need to think about money . . . unless it stops working.There is no value in the money itself, but it allows you to buy goods or services.

In the 1970s, we saw the world begin to adopt digital currencies. When people call Bitcoin a "digital currency," they're missing the point. The euro is a digital currency. The US dollar is a digital currency.

Less than three percent of these currencies exist in physical form; the rest is bits on ledgers. But the fundamental difference is that these ledgers are controlled by centralized organizations, whereas in Bitcoin, they're not. Bitcoin has a decentralized, open network.

Bitcoin is a small component of a digital currency. It's a cryptocurrency. It's a network-centric money.

A network that allows you to replace trust in institutions, trust in hierarchies, with trust on the network. The network acts as an arbiter of truth, resolving any disagreements about transactions and security in a way where no one has control.

The digital dream started for governments—the dream of being one day controlling every financial transaction of every human being on the planet in a way that everything was visible to the power structures.

This, of course, is the direction of where privacy dies. Where you can be scrutinized for every transaction you make. It would make it a complete system of global financial surveillance.

Paper bills were first used by the Chinese, who started carrying folding money during the Tang Dynasty (A.D. 618-907)—mostly in the form of privately issued bills of credit or exchange notes—and used it for more than five hundred years before the practice began to catch on in Europe and North America in the 17th century.

Paper money has been around for four hundred years. What were the shockwaves felt when the new form of money, checks and travellers' checks, appeared? Then, another major change to plastic cards. Major transformations in money have been seen before, but

Bitcoin is a radical transformation—even more so than the change from precious metals to paper money.

The New Invention

The next soon-to-be historical phase of money, Bitcoin, is an experiential technology. Bitcoin will evolve, and Blockchain will evolve alongside it.

To overstate the obvious, money today is what we earn and what we use to buy with. It's paper, plastic, and/or digital.

Bitcoin: A New Form of Money—What is the Nature of Money?

We use it as the basis of social interaction. Money is one of the least-studied technologies. Think about all the financial courses in the public-school system that were taught: none. In the majority of countries, money is not discussed in the education system. We don't discuss money, despite the fact that it affects every social and business interaction.

Do you store physical money in the bank? Unless it is in a safety deposit box, you have basically loaned it to the bank at one or two percent, and they loan it out to others in five and tenfold percentage.

Money as a Service

Cryptocurrencies are distributed digital currencies, much like a form of digital cash. They allow for seamless, direct, and extremely fast transactions between parties. You have full control over your payments and balance, and can spend and earn with ease and transparency.

Units of cryptocurrencies, often referred to as "coins", are stored in digital "wallets" that you use to manage your payments. Your wallet is protected by a complex password (private key) that only you know about. You can spend or send money by submitting a transaction from your wallet to someone else's using a "public address". Think of your wallet like a personal bank account that you maintain complete control over. We didn't need a third person there to help us make the transfer. We didn't need anyone to confirm that the coin went from me to you.

Now it's yours! I can't give you another coin because I don't have any left. I can't control it anymore. The coin left my possession completely. You have full control over that now. You can give it to your friend if you want, and then that friend can give it to his friend, and so on.

Problems

- What if more digital coins were created and added to one's balance whenever they wanted?
- How can I just hand over my digital coins to you in the usual way?

Solutions

What if we gave this ledger to everybody? Instead of the ledger living on a central computer or a server, it'll live in everybody's computers. All the transactions that have ever happened, from all time, in digital coins, will be recorded in it.

You can't cheat it. I can't send you digital coins I don't have, because then it wouldn't sync up with everybody else in the system. It'd be a tough system to beat. Especially if it got really big.

The Bitcoin protocol solves all apparent problems before they can arise.

What Does the Public Ledger Enable?

- Being open source, the total number of coins was defined in the public ledger at the beginning, that the exact amount exists. Within the system, they are limited.
- When an exchange is made, I now know that digital coin certifiably left my possession and is now completely yours. We couldn't say

that about digital things. Then, it will be updated and verified by the public ledger.

- Because it's a public ledger, I didn't need a third-party to make sure, I didn't cheat, or make extra copies for myself, or send twice, or thrice.

Within the system, the exchange of a digital coin is now just like the exchange of a physical one. It's now as good as seeing a physical coin leave my hand and dropped into your pocket. The exchange involved two people only. In other words, it behaves like a physical object. It's still digital.

We can now deal with one thousand coins, or one million or even .0000001 of a coin. It can be sent with the click of a button, and I can still drop it in your digital pocket if I was in London and you were all the way in New York.

The BitPay secure Bitcoin wallet for example, is giving consumers a powerful toolkit for getting started right with Bitcoin payments, and the BitPay Card gives Bitcoin users a fast way to convert Bitcoin into dollars and spend their funds anywhere Visa is accepted.

That is the power of innovation on the internet. It's innovation without permission. It's innovation without central approval. It's innovation without a broad network upgrade. And that means Bitcoin is not a specific financial network. It's not just a financial network for large transactions or small transactions, fast transactions or slow transactions. It's whatever you want to use it for, based upon what you choose to do at the endpoint.

What if the internet goes down? The same would happen to all credit cards, PayPal, banks, and other forms of money.

Each Bitcoin is worth the same as every other. Every Bitcoin wallet can confirm whether a Bitcoin is legitimate or not. Bitcoins cannot be counterfeited and they move across borders with ease. Moreover, they can't be blocked and don't require permission from a bank or 3rd party.

Bitcoin enables all kinds of economic activity without the need to trust anyone, and this is the future.

Why should we change? It all depends on your perspective. But it is innovation, progress that comes with blockbuster changes in how we do business around the world. Each year over 640 billion dollars have been sent globally to families 'back home'. If you are one of the many who does send money, then you can easily avoid the high expense and have it completed in minutes versus days. Think of migrant workers around the world sending money or small remittances back to families. What an impact! To send money today, you need to fill out forms, use a bank, have good reasons why the money is being sent, and the countries can still censor or block the transaction.

But even more powerful are the countries around the world like Venezuela and Zimbabwe, whose currencies are being hit by hyperinflation making the price of anything ten to one thousand-fold what it cost the previous day. Owing Bitcoin in these situations would be an extreme advantage, just for their survival. Even these governments are considering their own cryptocurrency like the 'petrodollar' from Venezuela. Bitcoin represents the perfect substitute.

5

PAYMENT OR E-PAYMENT?

Money is a system of control. Controlling money invokes power. The same is true of religion for the same reasons.

Children born today will not know how the world once functioned. Like the days before internet will be like the days we had cash and several credit and points cards. Soon, money will be a protocol. Most of the time, it can be really difficult to grasp the subject of 'Bitcoin', because it has many angles, such as computer science, maths, cryptography, economics, psychology, and sometimes philosophy too.

To really understand Bitcoin, one would need to understand much about decentralization, free-market economics, or intrinsic value.

The shadowy founder 'Satoshi Nakamoto' intended to keep the project totally decentralized without the possibility of human dependency or failure.

People can lose money if they are gambling or trading. Bitcoin markets are free markets, and this asset class is the most uncorrelated we have seen until now. The point is that Bitcoin is not backed by anything.

Do we know what backs the U.S. dollar? It used to be gold. Now it is a digital printing press. In August, 2019, the Federal Reserve 'printed' two hundred billion dollars to inject into the financial system—just a short-term measure. Bitcoin is backed by its maths, censorship resistance, and decentralization.

Cash

How did we get to be so lucky? How does it get better than this? This is a great time to be experiencing so many technological changes.

Cash usage is consistent across most U.S. household income levels and increases significantly at lower incomes. Thirty percent of U.S. households are unbanked or underbanked, with no or limited access to financial services.

Coins, Tokens & Altcoins

To answer how coins, tokens, and altcoins are related, we must first reflect back what cryptocurrencies are. At their core, cryptocurrencies are virtual currencies that are secured by cryptography. The underlying cryptographic technology uses encryption to ensure that units of currency cannot be duplicated, which facilitates the transfer of value. The various cryptocurrencies differ based on how a Blockchain's underlying asset is expected to be used and the design philosophy of the protocol.

Coins, tokens & altcoins, the three major types of cryptocurrencies, are differentiated based on their underlying technology and design uses per case.

For example, Bitcoin and other transactional cryptocurrencies are called cryptocurrency coins because of their use as a transfer of value.

By comparison, some are built on Ethereum and other competing Blockchains are referred to as tokens. Projects that are built on top of platforms like Ethereum do not themselves facilitate the transfer and are therefore merely tokens, not coins.

Finally, altcoins refer to most of the other projects that started off with Bitcoin's code but have now modified in some way.

Types of Reward Affinity Programs—Coins and Tokens

Air Mile points, Subway MyRewards, CSV Pharmacy, Shoppers Drugmart, Credit Cards, and endless other programs can be used for transactions, transferred, or converted into purchase of products or services. All these points are like a currency, have real value and in a sense part of the companies ecosystem designed for customer loyalty and branding. So, this ties in with organizations creating a new project opportunity that if you join you will be given a founder's opportunity to acquire some type of token relative to that project or community. As the project flourishes, the value of the token as an investment increases. Each type has various differentiators but all have a 'store of value'. Cryptocurrency coins that are used for transactions can still be further divided based on their uses and strengths. They can be converted into other cryptocurrencies instantly.

Value of Money

Value is neither created by money alone nor by trading. By itself, money doesn't do anything. Value is what happens when people create something for, build something for, or provide a service to other people.

Gamers, for example, have a long history of interacting with virtual goods and currencies in their virtual world. From Second Life to World of Warcraft to Fortnite, these virtual worlds are becoming multi-billion-dollar economies that bleed into real-world economies.

Gods Unchained is a Blockchain-based strategy trading card game that leverages the Ethereum Blockchain, by launching debit-card-enabled in-game purchases. In addition, to allowing payment for card packs in either, more players are allowed to participate, regard less of their Blockchain knowledge. In fact, they don't even need to know anything about a Blockchain other than it just works.

A Blockchain card doesn't just sit in a database; it has the ability to travel the world and know where it's been. eSports pros, streamers, and artists alike can digitally sign and sell digital items to fans.

In the words of Robbie Ferguson, Gods Unchained co-founder: "Blockchain isn't just for digital currency; it is laying the foundation for a whole new digital economy. Gods Unchained will become a game that any person can play, regardless of their Blockchain familiarity".

Money isn't real. It is a concept meant to create scarcity. Type it into your computer and into your bank account. NASA, for example, spends trillions, but these are just numbers. Money only has value because we choose to give it value. Money is not wealth.

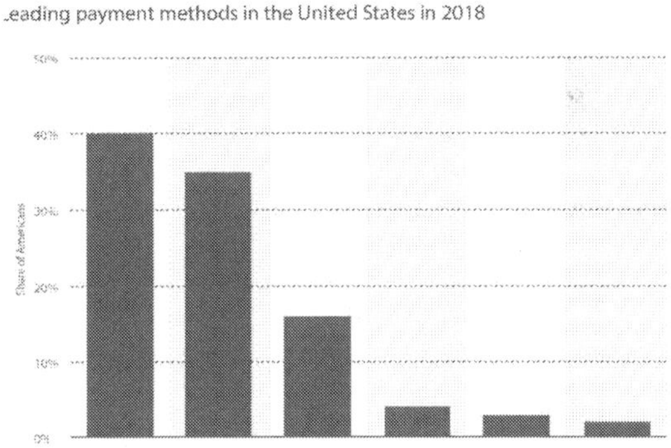

Preferred payment method in US from StatistaInc.

Mobile payments have grown from a niche service for tech-savvy consumers to mainstream use in the space of several years. The current trends in the mobile payment space will only increase as even more emerging technologies like cryptocurrencies will transform the world away from cash.

WeChat—One Billion Users

Overall, China is in an economic takeoff phase whereby the majority of the working population moves from agricultural jobs to technical jobs and the city. In the late 80's, about 24% of the population lived in cities, and now there are more than 50%, a figure that's expected to climb to 75% within ten years. The use of cash as we know it is rapidly disappearing and is being taken over by new payment systems made via mobile phone.

In 2012, a messaging app was developed by Internet giant Tencent, which has since evolved into what is described as China's "app for everything" and a "super app" because of its wide range of functions. The application WeChat Pay is a payment feature integrated into the WeChat app, which users can use to complete payments quickly with smartphones. Over one billion users have adopted the technology, and most of these users don't use cash at all. WeChat has Quick Pay, QR Code Payments, In-App Web-Based Payments, and Native In-App Payments, collectively to handle a wide range of payment situations. The machine-readable code (QR code) on a smart phone pays for everything. Tencent's embracement of this technology is stated in their vision to "Connect people to people, people to services, people to businesses, and people to objects."

M-Pesa

The African continent does not have the infrastructure to support credit cards wide scale. Introducing: M-Pesa (M for mobile, pesa is Swahili for money), a mobile phone-based money transfer financing, and microfinancing service, launched in 2007 by Vodafone for Safaricom and Vodacom, the largest mobile network operators in Kenya and Tanzania.

M-Pesa is more than a way to move money; it's effectively a bank, and it's introducing basic financial services to tens of millions that have never used them. The system has bank-level security to protect against money laundering, and because users are registered, there's a record of every transaction that includes first and last names and ID numbers. The system is so convenient and reliable that companies can pay their employees this way. The transaction numbers are stunning: in a country of forty-three million people, M-Pesa has over eighteen million registered users and has been so successful it has introduced the system in India, Albania, Egypt, Romania, and Eastern Europe—areas with large unbanked populations.

Ali-Pay

The creation of Ali-Pay in 2004 by the Alibaba Group in China is a third-party mobile and online payment platform with over 870 million users. Alipay overtook PayPal as the world's largest mobile payment platform. Alipay is very similar to Apple Pay, WeChat, and PayPal because it overlays traditional card payment methods. It's what you were use to experiencing in different apps, now all gathering into one place at blistering speed.

In the U.S., mobile bill payment has become a preferred method of banking for many Americans. In 2019, 47% of Americans using their mobile phones for at least some form of payment processing. Some retailers have integrated in-store mobile payments with loyalty programs, incentivizing customers to earn points every time they make a purchase. E-commerce platforms are gaining a larger share of retail spending each year, partly due to the lack of friction during the purchasing process. Part of that reward comes from the convenience of contactless payments, and part from discounts or coupons earned by visiting the store initially.

Privacy—'Free' Services from Google and Facebook

Only you are responsible for your digital privacy. The government clearly isn't going to protect you. Sure, the digital corporations sell ads and generate revenue. But the real currency here is your information.

Your name, age, gender, location, search habits, interests, frequently visited websites, shopping habits, estimated income: these are all commodities bought and sold in the back alleys of thc internet by the major tech companies. They call it free speech, so they can provide you with a better ad experience. They justify it by offering you "free" services. But you didn't actually think all those services were really free, did you?

Europe's highest court ruled in favor of Alphabet Inc.'s (Nasdaq: GOOG) Google. The case centered on Google users' "right to be forgotten."

At its core, "right to be forgotten" is your right as a private citizen to request that companies, such as Google and Facebook, "forget," or remove, any information about you that's "inadequate, irrelevant or no longer relevant, or excessive." In short, it means that you can ask the Googles of the world to remove your personal information or details of previous crimes from global search results.

Basically, the European Union ruled that if you ask Google to "forget" your information, it must remove it from European search engines only. The rest of the world is fair game.

The Takeaway

While Google hailed the ruling as a win for free speech, the costs Google would incur to remove that information globally, and the potential for lost revenue on search results would be huge.

Remember, it's always about money.

People should control and even own their own data, period! People ought to have the right to decide what, when, how, and how much about their identities are shared with others. Respecting one's right to privacy is not the same as respecting one's actual privacy. We need to do both. By eliminating the need to trust others, Bitcoin makes this possible.

In the cyber security world, I stated, "It's not a matter if you get breached, it's a matter of when." It's all about keeping our own information under our own control. Privacy to us is a basic human right, and it's the foundation of freedom and free societies. In the last 20 years of the Internet, central databases in both public and private sectors have accumulated all sorts of confidential information about individuals and institutions, sometimes without their knowledge.

The elusive Satoshi Nakamoto eliminated the need to know the true identities of people in order to interact with them. The problem this solves is huge. He installed no identity requirements for the Bitcoin Blockchain network layer itself. No one has to provide a name, an email address, or any other personal data in order to download and use the Bitcoin software. The Blockchain doesn't need to know who anybody actually is.

Corporations can create what we could call 'cyberclones' of people, the virtual you, by fracking the digital world for corresponding data. This 'virtual you' may know more about you than you do, because you can't remember what you did a year ago, or what you said a year ago, or bought a year ago, or your exact location, or what drugs you took, or what diagnosis was made by your doctor, or what test score you got.

The trouble is that the 'virtual you' would not be owned by you. The identity would be owned by large corporations. Even democratic governments are creating surveillance nations.

Credit cards are very identity-centric. Your name, your address, your financial information, your personal identification number, and your pin: they're all attached. That's why millions of people's addresses and phone numbers are stolen every time a database gets breached. What does it mean for a corporation to have perfect information about you?

Many people are simply unaware of how much privacy they bargain away every day, leaving a trail of digital footprints.

Blockchain protocols allow us to choose the level of privacy that we're comfortable with in any given transaction or environment. We can build high or low levels of transparency into any application, business model, or set of transactions, if the stakeholders agree that it's a good idea.

6

BLOCKCHAIN AGAINST THE HACKERS

Bitcoin keys can be stolen. Blockchain has been hacked. The configuration was poorly set up.

While it is almost impossible for Blockchain to be hacked, it has been. An initial set up of a Blockchain requires programming, and if there are flaws or bugs that are overlooked, they are not easily fixable like today's software using a 'software patch'.

In total, hackers have stolen nearly two billion dollars worth of cryptocurrency since the beginning of 2017, mostly from exchanges, and thats just what has been revealed publicly.

We've started to see what this means in practice and what these inherent weaknesses could mean for the future of Blockchains and digital assets.

In the cyber security world, organizations have been looking for "the silver bullet", a security software solution that really protects their critical data assets.

Today, large corporations can easily have in their security deployment one hundred different types of solutions fending off thousands of attacks from the bad guys who just need to find one crack in the system . . . which they so often do.

Overall, anyone can be hacked any time. Any time you connect to the internet with any device, there is the potential to be hacked. Anyone or any organization that has a network connected to the internet or provides an online service is open to an attack.

To control your Bitcoin, you need to control your private keys. Specially designed storage units called digital wallets can be safe, but many are leaving their private keys with a Cryptocurrency Exchange, and those Exchanges are getting hacked. Once your private keys are gone, there is no recourse. You are responsible.

Exchanges are like banks, in that they can facilitate transactions. They can store your assets and are responsible for the security for them along with credit card fraud issues. As long as you have done your reasonable due diligence, you are protected. But if Exchanges get hacked, your private keys/assets are gone with no reimbursement of any kind.

Who is Doing the Hacking?

Hackers come in many different flavours with many naming conventions: it's the Wild West vs Lawmakers.

In the world of Cyber Security, there are many groups.

WHITE Hat Hacker

So, there are the good guys and there are bad guys. A 'white hat hacker' is the good guy. These are ethical hackers who have usually gone through security training, though many are hobbyists.

Ethical hackers are the people who test existing internet infrastructures to research loopholes in the system. They create algorithms and perform multiple methodologies to break into systems, only to strengthen them.

The purpose of ethical hacking is to evaluate the security of and identify vulnerabilities in systems, networks or system infrastructure. This is often referred to as PENN (penetration) testing. It includes finding and attempting to exploit any vulnerability to determine whether unauthorized access or other malicious activities are possible. For example, when Microsoft does a software release the beta has been given to thousands to test out for bugs and other deficiencies.

GREY Hat Hacker

A 'grey hat hacker' has mixed intentions. This hacker type has neither the good intentions of a white hat hacker nor the ill intentions of a 'black hat hacker'. A grey hat would break into systems but never for his own benefit. Famous grey hat hackers have exploited systems only to make the information public, and to bring to limelight vast datasets of information that contains wrongdoings. For example, the company Ashley Madison (an extramarital affair service) was hacked by a grey hat because the company lied when they stated they did not keep personal information stored about ex-members. So, all the information on people having affairs ended up being exposed to the public.

BLACK Hat Hacker:

These are the bad guys that may actually help identify the weaknesses in all organizations' security but overall are in it to profit, either financial or politically. There are millions of bad guys all over the world, from the fourteen-year-old in the bedroom to the state-sponsored actor looking to steal identities or take down a power grid in some country. All the recent hacks of the Democratic Party, including Hilary Clinton's unsecured private server, were a dream come true for the black hats.

Black hats can even use a Hosted Solution with SAS (Software-as-a-Solution) for malicious attacks. This activity isn't confined to Russia. This allows the bad guys to use state-of-the-art services of a company that already has all the infrastructure in place. They pay a monthly subscription fee for any flavour of cyber attack they want to engage in. The host provides bandwidth and specific malware for specific targets, including phishing, spear-phishing, D-Dos, Brute Force attacks, and more. You can solicit these bad guys on the Dark Web.

An organization called AVtest.org estimates that there are 250,000 malicious programs created every day with a minimum number of 1.2 million black hats.

Ransomware is one of the most popular attacks made on individual computers, Fortune 500 companies, and government agencies. Malicious software encrypts all data on a system and then demands payment in Bitcoin for the de-encryption key to get their data back.

RED Hat Hacker

Another good hacker type to break the stereotype, the red hat hacker, acts ruthlessly towards black hat hackers. Their sole objective is to destroy the effort of every bad hacker type and to bring their entire infrastructure down. A red hat hacker would be on the lookout for a black hat hack, intercept it and hack into the black hat hacker's system. Not only would this halt the attack, but also push the black hat hacker out of business! Anonymous, for example, was a hacktivist group who hacked and exposed The Church of Scientology, the KKK, and even ISIS. Some companies use the red hats as a service to support their cyber security solutions.

Blockchain

If the bad actors can profit from attacking the Blockchain, they will. Bad guys have always been ahead of the curve. In the 1930's, Bonnie and Clyde stayed ahead of the curve by having the fastest getaway cars and machine guns that trumped law enforcement's puny pistols.

The black hats have already tried to hack the Blockchain but have failed. But powerful groups/and or state actors might team up in an attempt to exploit the system.

The Sybil attack on Blockchain

The name "Sybil attack" comes from a book, Sybil, about a woman with dissociative identity disorder. Microsoft researchers first investigated and wrote about the implications of Sybil attacks on peer networks in the early 2000's.

A Sybil attack uses multiple fake identities in an attempt to control a peer to peer network. Although they appear to be unique users behind the scenes, a single entity controls many identities at once, which can influence the network through additional voting power in a democratic network.

The United States' recent trouble with Russian election influence via fake accounts used bots & didn't hack into Facebook or Twitter, but they used multiple identities to influence the entire network. Facebook didn't even realize the extent of fake accounts on their platform until internal investigations months after much of the damage was already done.

A successful Sybil attack against a Blockchain or file transfer network would allow bad actors disproportionate control over the network and interrupt the flow of information across the network.

Preventing Sybil Attacks

There is however, a wall of obstacles that would prevent a successful attack or other to the Blockchain network. Significant technical and security changes need to happen as the world prepares for the Internet of Things. Not just your thermostat or fridge will be connected to the internet. Every appliance will hold a tiny microchip. Effectively, billions of devices around the world.

Blockchain has already moved from version 1.0 to 2.0 and needs to continue to evolve, adding more security and stability.

Cost to Create an Identity:A Chain of Trust

Blockchains use the cost of creation of identities on the network through mining. In proof of work algorithms, in order to create a new identity on the mining network, you'll need another computer with processing power to contribute. This attaches a significant cost to adding hundreds or thousands of pseudonymous nodes that might be able to influence the adoption of a major network change referred to as a fork or other Blockchain vote.

Two-factor Authenticated Security Code

The same goes for proof of stake, where purchasing computing power is replaced by staking currency. There's a cost to join the network and have a vote. That resource requirement limits the number of accounts a bad actor can create.

Chain of Trust

Some type of trust is needed before allowing a new identity to join the network. This usually takes the form of a reputation system, where only established, long-term users can invite or vouch for new entrants to the network. Chain of trust also extends to outright identity verification. An attacker would have to go against all the honest nodes and still produce a longer chain. As changing a block changes its hash and hence breaks the chain, this attacker must invest huge computing power in the case of a proof-of-work—especially when several blocks have been appended to the one he or she wants to change.

When enough blocks have been appended, an attacker must surpass the power of all other nodes, and this is considered to be impossible.

How do you know if you are getting hacked? Who is hacking you? Why are they hacking you? It is not good enough to protect your password anymore. We all should be learning more about security. Two-factor identification now should move to multifactor when you sign in somewhere. So many exchanges that hold cryptocurrencies have been hacked with 75% of them are failing at even at passwords.

What about the bad guys, security, and cash? What about money laundering?

Bad guys use cash perfectly well. After all, cash is the original, anonymous, peer-to-peer currency. Why do we need to oversee every transaction? The bad guys will do bad things anyway. But it turns out that millions of people have been punished by not being included in the banking system, cast (unduly) as possible criminals.

Remember, the banks too can be bad. When economies and systems go South, and the banks are themselves in a financial squeeze, they can turn on their customers. Amidst the 2008 financial crisis, the government and banks in Greece only allowed sixty euros to be withdrawn per day, and later 20% of citizens' pensions were taken in order to pay off the debt.

In North America, Pay Day loan franchises are a booming business, literally cashing in on the people that banks have excluded from their system. Some thirty million are unbanked, and the only way to cash a check, send money, or obtain small loans is to pay exorbitant fees. So many people then get trapped in the payday loan cycle that is almost impossible to get out.

In 2016 and 2017, the government in Greece was encouraging the use of credit cards or debit cards to pay for goods and services, in order to reduce cash-only payments. By January, 2017, taxpayers were only granted tax-allowances or deductions when payments were made electronically, with a "paper trail" of the transactions that the government could easily audit, in order to reduce the problem of businesses taking payments but not issuing an invoice—purportedly.

Financial Fraud and Creative Accounting—A Blurring of Lines

Blockchain will provide transparency here, cleaning up the grey areas of auditing. The acquisition of Wachovia(the 4th largest bank holding company in the US) by Wells Fargo was completed on December 31, 2008, after a government-forced sale to avoid Wachovia's failure. The Wachovia brand was absorbed into the Wells Fargo brand.

In March, 2010, Wachovia admitted to "serious and systemic" violations of the Bank Secrecy Act for laundering $378 billion between

2004 and 2007, the largest violation in terms of a dollar amount. It allowed Mexican and Colombian drug cartels to launder money through casas de cambio (money exchange houses) by willfully failing to set up an effective anti-money-laundering program.

Wells Fargo couldn't stop there. The Wells Fargo account fraud scandal is an ongoing controversy brought about by the creation of millions of fraudulent savings and checking accounts on behalf of Wells Fargo clients without their consent. In December of 2018, a settlement announcement indicated that Wells Fargo had already paid $2.3 billion in settlements and consent orders, so its $575 million settlement brought the total to nearly three billion.

In April, 2018, new allegations against Wells Fargo were reported, including signing unwitting customers up for unnecessary auto insurance policies, with the possibility of an additional billion-dollar fine. The company later paid this sum.

It seems that many companies around the world can 'cook their books' and manipulate their seeming profits, in order to present a favourable public outlook. The figures which are fed twice a year to the investing public have all been changed in order to protect the guilty. In fact, this deception is all in perfectly good taste. It is totally legitimate. It is called creative accounting. Once Blockchain is well in place some of these fraudulent practices can be reduced or eliminated.

Blockchain Credit Card

For all intensive purposes, Blockchain is like a new application built on an understanding of currencies' store of value.

In the second generation of the internet, it holds the potential to transform money, business, government and society.

Before, block and chain were used separately to explain the concept of many blocks, each containing data on transactions, all connected in a chain. Over the years Blockchain would become the standard term used for this technology.

In 1991, a paper written by Stuart Haber & W. Scott Stornetta entitled, "How to Time-Stamp a Digital Document", the concept of time stamping digital documents was proposed, to ensure transactions were 'signed' at a certain time.

Years followed with many enhancements with cryptography describing a method in which computers can prevent attackers from modifying or tampering with any previous logs made, while at the same time making any logs unreadable to the attacker.

From 2002 to 2007, more developments occurred. For one, a network file system with a decentralized trust was proposed. This was a proto-Blockchain, in that the writers to this file system would trust one another but not the system itself. All of this lead up to 2008 with Satoshi Nakamoto. To this day, Blockchain has continued to evolve, and now Blockchain 2.0 has emerged. With it comes open source capability for anyone to write software on top of the Blockchain.

A Blockchain, originally block chain, is a growing list of records, called 'blocks', that are linked using cryptography. Each block contains a cryptographic hash of the previous block, a timestamp, and transaction data.

By design, a Blockchain is resistant to modification of the data. It is an open, distributed ledger that can record transactions between two parties efficiently and in a verifiable and permanent way.

For use as a distributed ledger, a Blockchain is typically managed by a peer-to-peer network collectively adhering to a protocol for inter-node communication and validating new blocks.

Once recorded, the data in any given block cannot be altered retroactively without alteration of all subsequent blocks, which requires consensus of the network majority.

There are mainly three types of Blockchains that have emerged after Bitcoin introduced Blockchain to the world.

- Public Blockchain
- Private Blockchain
- Consortium or Federated Blockchain

Public Blockchain

A public Blockchain, as its name suggests, is the Blockchain of the public, meaning a kind of Blockchain which is'for the people, by the people and of the people'.

Here no one is in charge and anyone can participate in reading/writing/auditing the Blockchain. Another thing is that these types of Blockchain are open and transparent, so anyone can review anything at a given point of time on a public Blockchain.

But a natural question that comes to our mind when no one is in charge is this: "How are the decisions being made (with Blockchain) exactly?" The answer is that decision-making happens by various decentralized consensus mechanisms such as proof of work (POW) and proof of stake(POS) etc.

Private Blockchain

Private Blockchain, as its name suggests, is the private property of an individual or an organization. Unlike public Blockchain, there is a party in charge who looks after all of the important things such as read/write and whom to selectively give access to.

Here, the decision is left to the discretion of the central in-charge who can give mining rights to anyone, or not give at all. That's what makes it centralized again, where various rights are exercised and vested in a central, trusted party but, yet, it is cryptographical, secured from the company's point of view and more cost-effective for them. But it is still debatable whether such a private thing can be called a 'Blockchain', because it fundamentally defeats the whole purpose of Blockchain that Bitcoin introduced us to.

Consortium or Federated Blockchain

This type of Blockchain tries to remove the sole autonomy which gets vested in just one entity by using private Blockchains. Here, instead of one entity being in charge, you have more than one party. Basically, you have a group of companies or representative individuals coming together and making decisions for the best benefit of the whole network. Such groups are also called consortiums or federations—hence the name.

For example, let's suppose you have a consortium of the world's top-twenty financial institutes. If a transaction, block, or decision is voted/verified by more than, say, fifteen institutes, only then would it be added to the Blockchain.

So, it is a way of achieving things much faster, and you also have more than one single point of possible failure, which in a way protects the whole ecosystem.

We require more types of Blockchain for the following reasons:

- One no longer needs to rely upon huge servers.
- They are cost effective and fast.
- They reduce the need for more trusted parties, because you can implement smart contracts instead of them.
- Gives options for rights and access management while leveraging the same Blockchain technology and reaping its benefits.
- Reduces redundant work.
- Distributed consensus between interested parties becomes fast even though you are geographically segregated.

Banks are centralized, meaning they can be hacked. They slow things down. They charge too much. They exclude people from financial services.

What if there wasn't just an internet of information? What if there was an internet of value? What if we implemented a vast, globally distributed ledger?

This is the paper Satoshi Nakamota wrote on this.

Trust is not possible through intermediary organizations—only through cryptography by collaboration and clever code.

The first miner to gain consensus is paid some of the cryptocurrency from the Blockchain. Then, that block is connected to the previous block to verify the fact that all transactions are real.

So, Bitcoin itself is a cryptocurrency. Ethereum is very interesting because, unlike Bitcoin which has a single function as currency, Ethereum allows the development of any application and is significant to the financial service industry.

Think of email as the first big app of the internet. And Bitcoin is the first big app of the Blockchain. On Ethereum, you can build Smart Contracts—contracts that self-execute and self-police. Inside, it is like having a bank, a lawyer, and a system of government.

Blockchain—The Trust Machine

So, with the internet of information, we have to rely on powerful intermediaries to establish trust. Banks, governments, PayPal, and even social media companies like Facebook work to establish our identity and ownership of assets. They help us transfer value and settle transactions.

Great job, but . . .

- They use centralized servers, which can be hacked.
- They take a fee for their services.
- They capture our data, not just preventing us from monetizing it, but often undermining our privacy.
- They are sometimes unreliable and often slow.
- They exclude two billion people who don't have enough money to justify a bank account. Thirty million of them are in the US

Enter the Blockchain. Its protocol establishes the rules—in the form of globally distributed computations along with encryption—that ensure the integrity of the data traded among billions of devices, without going through a trusted third party.

Trust is hard-coded into the platform. It acts as a ledger of accounts, a database, a notary, a sentry, and clearing house, all by consensus.

Why Should You Care?

Every business, institution, government and individual can benefit in profound ways. The Blockchain is already disrupting the financial services industry. How about these billions of connected smart things that will be sensing, responding, sharing data, generating, and trading their own electricity, protecting our environment, managing our homes and our health?

And this Internet of Everything will need a Ledger of Everything. And how about growing social inequality? How about redistribution of wealth? As with all major paradigm shifts, there will be winners and losers. But, if we do this right, Blockchain technology can mean prosperity for all.

A Blockchain is a global, online database that anyone with an internet connection can use. Unlike traditional databases—which are typically owned by big institutions—a Blockchain doesn't belong to anyone, and with an entire network of people monitoring it, cheating the system by faking documents, transactions, and other information is nearly impossible. Blockchain stores information permanently across a network of personal computers. It not only decentralised the information but distributes it too.

So, how does an online multi-purpose database work for everyday use? How does it stay relatively hack-proof?

The answer is Blockchain's millions of users. They make it difficult for any one person to take down the network or corrupt it. The people use their own personal records to hold bundles of records submitted by others. The records are known as blocks. Each block has a timestamp and a link to a previous block forming a chronological chain. You can view it and add to it, but you can't change the information that is already there.

This means records can't be counterfeited or altered by someone else.

Blockchain's main application is Bitcoin, a digital currency that is created and held electronically and can be sent to anyone, whether you know them or not. Unlike credit cards and PayPal payments, there are no middlemen, like the banks, asking for personal information. Instead, people all over the world move the digital money by validating the digital money used in other users' Bitcoin transactions, earning a small fee in the process. Where the Blockchain comes in verifying the ownership of this digital cash and making sure only one person is claiming it as their own at a time.

In 2017, a Deloitte study found that eight of the world's largest banks could save between eight and twelve billion a year by using Blockchain.

7

TOKENS AND MYTHS

Bitcoin Myths

This is what actually happens throughout history when you introduce a disruptive technology. You meet resistance. Resistance is the first reaction.

People are often suspicious of anything new, especially new things that aren't easy to understand. Ten years from now, Bitcoin and Blockchain will be well known widely used, just like the internet is today. But trying to understand it will be very confusing and a long learning curve lies ahead.

Bitcoin was Hacked

What has been hacked is the central systems that have used Bitcoin. Exchanges and personal wallets have been hacked and Bitcoin has been stolen.

Bitcoin is Used to Extort People

Because of the semi-anonymous nature of Bitcoin, it has been and still is used in Ransomware attacks. Hackers actually breach networks and hold them hostage by encrypting the data and holding the encryption key until payment is made by Bitcoin. Generous instructions are given, and the majority of times, the victim is given the encryption key with instructions about decrypting the data.

Bitcoin is a Pyramid Scheme

Bitcoin is like a pure market. Its price fluctuates based on market supply, demand and perceived value.

Bitcoin will Collapse after 21 Million Coins are Mined: Bitcoin has a coded limit on the number of tokens it will release. The estimated date of Bitcoin issuing its last coin is believed to be in the year 2140. It might collapse. No one can predict what will happen at that point.

Enough Computing Power Can Take Over the Network: Possible. But many safeguards have been built into the system, and the payoff for the attacker is limited.

Virtually every aspect of how we live, move around, communicate, and work is embedded with immense quantities of energy using up a lot of electricity. Now it's even in our money. Bitcoin is energy-intensive, being used by so many nodes in the network consuming computer power to solve the mathematical puzzle for the reward of Bitcoin.

Many believe this is one of the biggest concerns: that Bitcoin could impact climate change. Some believe that Bitcoin is too energy-hungry and that the requisite energy could perhaps be redirected to help the billions of people who still can't get access to enough energy for the simple things in life, like the power to run a refrigerator, start a business, or create a decent job.

The California Energy Commission's energy comparison figures show an interesting story—the one about real people. It demonstrates a mind-blowing chasm of global energy inequality. California gamers will soon use more electricity than one hundred million people in Ethiopia. Pools and hot tubs in California use more energy than the entire island of Jamaica.

In all emerging economies the cost and reliability of electricity are still the leading constraints to economic growth and job creation. The climate versus bitcoin debate is a distraction. The bigger question is why these countries don't have access to more energy. Even more to the point, what can be done to solve this problem?

Bitcoin is a Good Investment

It is an unstable investment. It is extremely speculative and only worth what people are willing to trade for it. The hope for many willing to risk a lot of their capital is for Bitcoin to skyrocket.

Blockchain Truths: There's no doubt about the potential that the Blockchain technology has to offer. The world has accepted it at large. Blockchain developers are solving multiple problems across industries with this technology.

But there's another side to the story wherein the hype is shadowing real implementation issues with the Blockchain. These issues hinder the mass-adoption of this world-changing technology.

There's no doubt that we'll eventually get to the point where we've fully explored this great technology. However, to get there, it's essential to face realities, both as developers and enthusiasts of the Blockchain.

Limited Scalability

One of the key advantages of the Blockchain technology is decentralization. However, it comes at the cost scalability. As the number of transactions increase (and since every transaction passes through each block), the requirement increases too. Eventually, it's not possible for every node to participate, and the resources to process each block then rests with a limited few. However, that potentially compromises the decentralization aspect of the Blockchain. Scalability is probably the number one barrier when it comes to Blockchain adaptability. The Lightning Network might just evolve as a full solution here.

Storage Constraints

Almost every application that's built on the Blockchain will require some form of storage. It can be either to store user identities or to record specific financial as well legal data. However, by property traits, every node stores the data. Along with it, no data will be able to be removed, regardless of whether it's needed or not. The above two cases put a considerable cost on the storage operation when it comes to the Blockchain. Thus, storage constraints are a massive hurdle that stands in the way of applications being build on the Blockchain.

Lack of Standards and Governance

No central authority governs public Blockchain. While that's great to build a trustless and open system, it has its disadvantages as well. For example, there's no one responsible for maintaining and upgrading the systems in the most efficient manner possible. Even if some developers band up to create an organization, it leads to some form of centralization (take, for instance, the Ethereum foundation). However, we can't even leave it completely open, as over the years it's proven ineffective on multiple levels.

Lack of Development Software

For any developer to work effectively and efficiently, the first requirement is adequate tooling. If inadequate, the developers can't do much, even with the best of intentions. As a matter of fact, the currently available pool of tooling is not good enough, even for the most seasoned of developers. Right from good testing frameworks to security auditing and even the right Blockchain analytical tool, the available tooling to a Blockchain developer is severely limited. Along with hindering the potential of the technology, it also demotivates enthusiasts.

Unfortunately, the hype makes Blockchain appear invincible, when the fact is that there's a lot of work left to be done. So, instead of making the technology glamorous, we need to focus on how to solve core implementation issues with the Blockchain.

All data and belief is that Blockchain is here to stay. However, the issues mentioned above (and many more) are preventing the technology from becoming mainstream. It's up to the developers, the investors, the Blockchain enthusiasts—to work towards a unified vision.

Trust

Trust in business is the expectation that the other party will act with integrity, which comes down to four values: honesty, consideration, accountability, and transparency.

So, we've come to rely on third parties to record transactions and do everything behind the scenes to make online commerce work. These intermediaries can be banks, governments, or businesses like PayPal and Visa.

Accountability means making clear commitments and sticking with them. Individuals and organizations must show they keep their promises and own up to any that they've broken.

Everyone involved should be able to verify this, and outside experts should be able to do so as well. Transparency means operating out in the open.

Although technology is the most trusted industry, social media companies like Facebook and Twitter have been used to manipulate public opinion.

Traditionally, trust in transactions—buying and selling goods and services—came from individuals or groups acting with integrity. We did business face-to-face with people we knew. Now, we often don't know who we're doing business with online, let alone whether or not people have integrity.

The Technical Perspective

Now with Blockchain, the network and code have built in new levels of trust and don't need to be established only by a third party. It means all parties in the system, not just the ones involved in the transaction, come to an agreement on what the facts are.

And once they agree and have consensus a new block is added. The transaction becomes a solid fact and part of the Blockchain's history. The ledger itself is what the trust is built on.

This new, solid foundation of trust will pave the way for citizens to have significant transparency of government organizations and tax distribution.

The transparency and trust built into Blockchain will start to rebuild the trust in our institutions.

What is relevant to understand is the underlying technology. Because of Blockchain, anything from a pokemon card to a traditional equity to an autonomous car can have unique digital presence, engage in commerce, and be traded in capital markets. If generative DNA can be stored on the Blockchain, and effectively managed to create over 100,000 unique individuals, it's not that big of a stretch to imagine medical data or personal identity taking a similar path. And all of this is happening right now.

Miners

Miners are people worldwide running "mining" software that confirms new blocks of recent transactions into the permanent Blockchain. In order to incentivize people to run the mining the Miners do work to find out the truth about which is block holds all the transactions in the last ten minutes. The miner uses their computing power to validate the block. The first miner to gain consensus is paid some of the cryptocurrency from the Blockchain. Then, that block is connected to the previous block to verify the fact that all transactions are real software, they are given a prize for their work—either in new coins, transaction fees, or both.

Anyone can see the new, unconfirmed transactions in real time. It's similar to gold mining, in that the Bitcoins exist in the protocol's design (just as the gold exists underground), but they haven't been brought out into the light yet (just as the gold hasn't yet been dug up).

The Bitcoin protocol stipulates that twenty-one million Bitcoins will exist at some point. What "miners" do is bring them out into the light, a few at a time.They get to do this as a reward for creating blocks of validated transactions and including them in the Blockchain.

Nodes

Backtracking a bit, let's talk about "nodes." A node is a powerful computer that runs the Bitcoin software and helps to keep Bitcoin running by participating in the relay of information. Anyone can run a node; you just download the Bitcoin software and leave a certain port open.

Nodes spread Bitcoin transactions around the network. One node will send information to a few nodes that it knows, who will relay the information to nodes that they know, etc. That way it ends up getting around the whole network pretty quickly.

Some nodes are mining nodes (usually referred to as "miners"). These group outstanding transactions into blocks and add them to the Blockchain by solving a complex mathematical puzzle that is part of the Bitcoin program, and including the answer in the block. The puzzle that needs solving is to find a number that, when combined with the data in the block and passed through a hash function (hash refers to a way to map data converting one value to another), produces a result that is within a certain range. This is much harder than it sounds.

Solving the Puzzle

How do they find this number? By guessing at random. So, that explains it or you can read a light technical version of it. The hash function makes it impossible to predict what the output will be. So, miners guess the mystery number and apply the hash function to the combination of that guessed number and the data in the block. The resulting hash has to start with a pre-established number of zeroes. There's no way of knowing which number will work, because two consecutive integers will give wildly varying results. What's more, there may be several arbitrary numbers or integers that produce the desired result, or there may be none (in which case the miners keep trying, but with a different block configuration).

The first miner to get a resulting hash within the desired range announces its victory to the rest of the network. All the other miners immediately stop work on that block and start trying to figure out the mystery number for the next one.

Difficulty

The difficulty of the calculation (the required number of zeroes at the beginning of the hash string) is adjusted frequently, so that it takes on average about ten minutes to process a block.

Why ten minutes? That is the amount of time that the Bitcoin developers think is necessary for a steady and diminishing flow of new coins until the maximum number of twenty-one million is reached (expected some time in 2140).

The broad outline of the genius of the programming and the concept:- for the first time, we have a system that allows for convenient digital transfers in a decentralized, trust-free and tamper-proof way.

This is where the Blockchain works behind the scenes. The computers connected to the Blockchain network are known as nodes. The nodes on the Blockchain network donate their processing power to solve a math problem associated with a new block. Although, the Bitcoin community calls these people "miners," but it's actually the computers doing all the work. There is no skill involved.

Each miner (person/node) runs the Blockchain software like a utility function in the background of their computers (while others use high grade equipment) and specialized computers to boost their efficiency, but anyone with a computer can do it.

The equipment, the software makes mining a level playing field. The math problem itself is untouched by human hands. Not all nodes are mining, however. Most of the nodes on the Bitcoin network simply verify data. They send it to peer connection.

Miners gather up the recent transactions, order them into a block of data, and add it to the chain. Each block must refer to the previous one to be valid. And since the Blockchain is public, Bitcoin is more traceable than cash. When the network reaches consensus about what happened, it records it on the Blockchain.

How the network reaches consensus is a critical factor. The Bitcoin network relies on what is called a proof of work. Because we can't rely on the identity of the miners to select who creates the next block, the network instead creates a puzzle. Miners use their resources, namely computing hardware and electricity, to solve the puzzle by finding the right hash. Remember a hash is a unique fingerprint for the text or the data file. This puzzle requires a lot of computing to solve. But when someone solves it, someone else can check the work quickly. Whoever solves the problem first gets to create the next block. And for each block a miner creates, he or she receives Bitcoin as a reward. There are no shortcuts to solving the problem, so when the rest of the network sees the answer, everyone trusts a lot of work went into producing it. On average, a miner finds a hash meeting the target every ten minutes. Sometimes it takes longer, sometimes it doesn't take much time at all.

The various cryptocurrencies differ based on how a Blockchain's underlying asset is expected to be used and the design philosophy of the protocol.

Transactional Coins

These are the frontrunners for transactional coins at the moment, because they offer near-instant, feeless transactions which are orders of magnitude faster than that of Bitcoin. Both of these could facilitate peer to peer and machine to machine transactions in the future. Bitcoin has the first-mover advantage in the cryptocurrency world and also boasts of having the longest running Blockchain.

There is actually a cryptocurrency named Dai which is price-stabilized agianst the value of the US Dollar.

Privacy

There is a category of coins called private coins. The leader is Monero and is a cryptocurrency that hides data about its users. It uses stealth addresses and other privacy-enhancing features.At a minimum, privacy coins hide identities. They also often hide the amount of cryptocurency traded and held in wallets.

Types of Tokens

Tokens are not a cryptocurrency but are used for platforms based on their use case and structure. For example, think of a company that creates an amusement park and offers tokens at a low price before it opens allowing you to use all the services inside. If you wait to buy the tokens after opening the cost will be much higher.

Work Token

Work tokens are required to get rights to participate in a platform. The functionality of this token will expand over time in step with the adoption, number of users and growth of the project .

Hybrid Token

These are tokens that have traits of different tokens that allows them to unlock features of a network and acting as a means of exchange to interoperate with other networks.

Network Token

This type is tied to the value and development of a network. They are used to pay for transactions on the network, is a network token, as it is only usable when the network grows.

Equity Token

These are tokens that behave similarly to a company stock, and are used to vote, share dividends, and make decisions. With these tokens, holders are considered shareholders.

Smart Contracts

A Smart Contract is just like how Blockchain functions, but with the pre-programmed terms of a contract. Execution of the contract happens on its own, when both the parties fulfill the conditions laid out. The Smart Contract program, upon agreement of price between the two parties, executes the contract and performs the transaction automatically.

The software verifies the owner, performs the transaction, and ships the product or transfers ownership registrations on its own. Smart Contracts can execute purchase or sale of the house, cars, financial instruments, marriage registration, divorce, property registrations, etc., without human intervention.

The benefit is that it eliminates intermediaries like banks, brokers, and agents, and the individual performs the entire transaction on his or her own.

The key use cases or application of Blockchain technology will multiply in the coming years. It will take over most of the processes of many industries all over the world. It is also set to become an integral part of human lives in all our day-to-day activities.

Smart Contracts and the Preservation of Rights

Rights must be protected. Ownership rights should be transparent, and they should be enforceable. Individual freedoms need to be recognized and respected. Smart contracts are an instrument for protecting, and enforcing those rights. The problem that this principle solves is preserving our ownership rights without impinging on our freedoms.

As the Internet became a venue for the presentation of all types of content, photographs, music, and movies, we adapted laws and standards developed for the physical space in the online world.

In this migration of existing laws into online practices, legitimate rights got trampled. The rights not only to privacy and security, but also free speech, reputation, and equal participation.

People could anonymously censor at will (at little cost to themselves). Artists and filmmakers saw revenue streams dry up, as their fans uploaded digital files for others to download for free.

In 2015, Grammy winner recording star Imogen Heap partnered with Ujo company, who use the Ethereum platform, eliminating the confusion of music ownership by paying artists using smart contracts and cryptocurrencies. Thereby having royalties go to whom they should. The first recording artist to do this,

In 2017, Spotify acquired Media Chain, a peer-to-peer Blockchain database for sharing information across different applications and organizations. The company issues Smart Contracts to musicians which will ensure artists are paid appropriate royalties.

Sony, Netflix, and YouTube are among the many exploring the use of Blockchain to help identify the rightful music rights holders and originators, so they can receive fair royalty payments.

The Blockchain breakthrough is strengthening property rights by making them inarguable and unforgeable.

It's predicted that Smart Contracts will become more and more common, as we learn how to move our code of laws into computer code and let the software determine the legal course of action.

It goes much beyond technology, physical assets, intellectual property, and privacy tools. A deeper understanding of rights and management systems needs to be facilitated.

8

GLOBAL ADOPTION OF BITCOIN

Baby Boomers: 1946 - 1964
Generation X: 1965 – 1980
Xennials: 1975 – 1985
Millennials (Gen Y): 1981 – 1996
Post-Millennials: 1997 – Present

Generation Z

Gen Z: Gen Z is the newest generation to be named. Members were born between 1995 and 2015. From 2010 onwards, Alpha kids have grown up with iPads in hand, never living without a smartphone or the ability to transfer a thought online in seconds. These massive technological changes, among others, make Generation Alpha the most transformative generation ever.

In July 1999, Robert Mundell an American who received a Nobel Memorial Prize in Economic Sciences monetary history and theory stated:

"I think that the internet is going to be one of the major forces for reducing the role of government. One thing that is missing but will soon be developed is the reliable e-cash, a method whereby on the

intent where you can transfer funds from A to B with either not knowing each other."

McDonald's Switzerland is Advertising with Bitcoin

McDonald's isn't just a fast-food chain—it's a thirty-billion-dollar real-estate company. But rather than collect a lot in royalties or sell its franchisees' cooking equipment and leverage it's supply chain, McDonald's makes much of its revenue by buying the physical properties and then leasing them to franchisees, often at large markups. In Switzerland, they are presently assessing moving to accept Bitcoin as seen by their advertisement

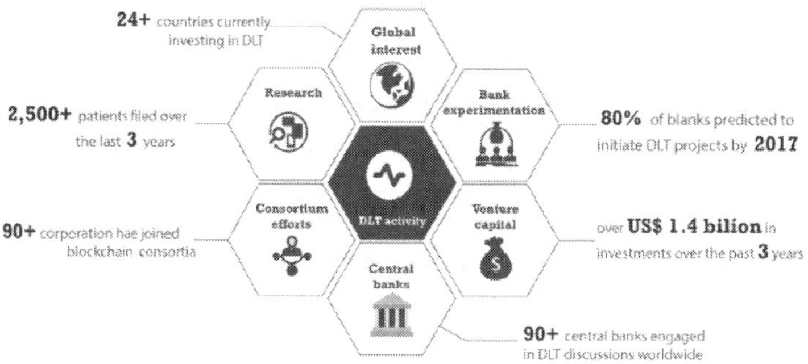

Burger King has been interactive in trial and errors with Bitcoin since 2016 in Europe, Russia, and other countries. The presence of Bitcoin's acceptance in the fast food industry will create awareness with millennial's and GenX being their major target market and fast forward the usage for crypto. It is now coming to the point for businesses to who is not rather than who is - planning or moving forward with Distributed Ledger Technology. Similar to today's new startups in Silicon Valley, the majority of companies use a Hosted solution that provides a complete infrastructure like a plug and play application in order to launch their business. These Hosted solutions will flourish because of the lack of Blockchain developers.

Microsoft accepts Bitcoin into your Microsoft account, you can use it to purchase games, movies, and apps in the Windows and Xbox stores. Not their online store. Amazon does not accept directly but use a third party.

How Many People Use Bitcoin in 2019? Facts and figures

An estimated 5% of Americans hold Bitcoin.
There are 7.1 million active Bitcoin users.
Leading exchange Coinbase has over thirteen million users.
Emerging markets users, who are often not considered in statistics, are likely in the millions.
Let's unpack those numbers a bit.

Millions of Bitcoin Wallets

According to Statista Inc. a leading statistics company: Forty-two million Blockchain wallets are in existence. Bitcoin users have several Bitcoin wallets and use multiple wallet addresses to increase their financial privacy when transacting in Bitcoin. Hence, the total number of Bitcoin users must be less than the number of wallets.

Percentage of Active Users

The number of people who use Bitcoin actively has also increased. A report from the leading Blockchain analysis company, Chainalysis, has revealed that active Bitcoin users make up around a third of the total number of Bitcoin owners. The rest are just holding.

The study used data collected until the month of August 2018. The researchers found that 2.3 million people use Bitcoin to make payments while 4.8 million users will hold Bitcoin to facilitate speculative trade of the cryptocurrency to turn a profit.

Those who use Bitcoin for payments come up to 11 percent of the total Bitcoin users, while speculators make up 23 percent. While these two groups actively move Bitcoin from their wallets almost daily, they do so for different reasons. However, the number of active Bitcoin users continues to grow.

The report also found that the number of long-term investors, also known as "HODLers," has grown as well. This group is the largest in size, coming up to thirty percent of total Bitcoin users. These are 6.4 million people who are holding Bitcoin for the long-term, in anticipation of gains to be made over the years.

Percentage of Americans Holding Bitcoin

A survey conducted by the Global Blockchain Council and SurveyMonkey that was taken by five thousand Americans found that sixty percent of Americans have heard of Bitcoin and that around five percent hold the digital currency.

If you extrapolate that to the entire U.S. population, that would mean that an estimated sixteen million Americans have invested in Bitcoin.

Coinbase's 13 Million+ Users

The most popular Bitcoin wallet and exchange provider, Coinbase, reportedly has over thirteen million users, which would suggest that the number of Bitcoin users is somewhere between thirteen million and thirty-two million.

Given that Coinbase is available in thirty-two countries and that it has emerged as the go-to Bitcoin wallet for newcomers, its user base number gives a better indication of how many users there are than purely analyzing wallet addresses.

In light of the limited data and the impossibility of accurately determining the number of Bitcoin users around the world, we can safely estimate that there will be around twenty-five million users globally. This number is after wallets not in use, and that many occupy several wallets

The World's first Bitcoin ATM

On October 29, 2013, a Robocoin machine opened in the Waves coffee shop in downtown Vancouver, Canada. This machine is understood to be the world's first publicly available Bitcoin ATM.

Now there are more than ten thousand around the world. The Bitcoin kiosks are machines which are connected to the Internet not a bank account but instead connect the user directly to a Bitcoin exchange. You put cash in for the exchange of bitcoins, given as a paper receipt or by moving money to a public key on the blockchain. The website *https://coinatmradar.com/countries* lists some the ATMs around the globe.

The Big Four Adoption

In March 2017, Deloitte, one of the largest accounting firms in the world announced that Deloitte Canada Toronto HQ office's internal restaurant called Bistro 1858, now accepts bitcoin. Oris Valiente, an original founder, and strategy leader of the Rubix by Deloitte team said "The first step toward wider blockchain adoption is knowledge, and the logical entry point is to first understand how bitcoin works. The goal is to make bitcoin more accessible so that users can experience blockchain technology first hand to deepen their understanding and ultimately increase adoption." "It's the best way to illustrate the power of the technology when you have a client from a different country and you can send them the currency without going through any banks, and there's a clear recognition that there was no connective intermediary in the process."

Fellow "Big Four" auditing firm PricewaterhouseCoopers has made similar supportive comments about Bitcoin in public, saying that using the currency is key to understanding its disruptive potential, rather than viewing it as a threat to the financial world.

In Nov 2017 PricewaterhouseCoopers (PWC), one of the world's largest accounting firms, accepted its first ever bitcoin payment for its advisory services. PwC is the second 'big four' accounting firm to accept bitcoin as payments for its services. Ernst & Young (EY) in Switzerland became the first firm of its kind to accept bitcoin for both auditing and advisory services since the turn of 2017. EY also installed a Bitcoin ATM among its offices, providing its employees with a digital bitcoin wallet within its in-house EY wallet app. Accepting bitcoin, EY believed was essential for their strategy of "digitizing itself".

Bitcoin User Growth in Emerging Markets

Bitcoin trading volumes from peer-to-peer exchanges show Bitcoin adoption is on the rise in economically troubled South American nations such as Venezuela, Brazil, and Colombia, as well as in African nations.

Who are the richest people in Bitcoin?

Cameron and Tyler Winklevoss.

Robert Parker, CEO of Holborn Assets in Dubai.

There are over fourteen million Bitcoins out there. According to some sources, there is only one holder that exceeds one hundred thousand Bitcoins.

However, the unknown inventor, Satoshi Nakamoto, is estimated to have mined one million Bitcoins in the currency's early days. The stash is supposedly stored across many wallets. So, Satoshi may be the richest person in Bitcoin.

The second largest Bitcoin wallet belongs to the US government—namely, the FBI.

In September 2019, the FBI shut down the Silk Road online drug marketplace, and it started seizing Bitcoins belonging to the Dread Pirate Roberts—the operator of the illicit online marketplace, who they say is an American man named Ross Ulbricht.

The seizure sparked an ongoing public discussion about the future of Bitcoin, the world's most popular digital currency, but it had an unforeseen side-effect: It made the FBI the holder of the world's biggest Bitcoin wallet.

The FBI now controls more than 144,000 Bitcoins that reside at a Bitcoin address that consolidates much of the seized Silk Road Bitcoins. Those 144,000 Bitcoins are worth close to $100 million at most recent exchange rates. Another address, containing Silk Road funds seized earlier by the FBI, contains nearly 30,000 Bitcoins ($20 million).

Who Owns the World's Biggest Bitcoin Wallet? The FBI!

Facebook Currency—Libra

The Libra Association is comprised of big-name companies such as Visa, Mastercard, PayPal, Uber and Spotify. However, Visa CEO and chairman, Alfred F. Kelly, recently said that no company has officially joined the project yet.

Facebook announced the Libra project in mid-June, and since then has faced intense regulatory scrutiny from around the world.

"Visa's continued interest in Libra stems from our belief that well-regulated Blockchain-based networks could extend the value of secure digital payments to a greater number of people and places, particularly in emerging and developing markets." — Visa spokesperson

"Visa has decided not to join the Libra Association at this time. We will continue to evaluate, and our ultimate decision will be determined by a number of factors, including the Association's ability to fully satisfy all requisite regulatory expectations," said Visa spokesperson. However, Visa said it still believes in the potential of Blockchain to be better than the existing payments system.

Mercado Pago, Visa, eBay, Stripe, and, Mastercard have become the latest five companies to officially depart from Facebook's Libra Stablecoin project, following PayPal, according to several news outlets, the firms themselves, and the Libra Association website. Mercado Pago is the payments arm of Mercado Libra, a Latin-American e-commerce giant. Libra has been a favorite target of world financial regulators since its announcement in June. European Central Bank members said it could destabilize the euro; and U.S. Congressional Representatives have called for an outright freeze on its development.

There are so many new reports, news feeds and daily flashes on so many announcements around the globe, here are a few.

August 23, 2019: Mark Carney, the current governor of the Bank of England, has offered a proposal that would replace the U.S. dollar with a digital currency similar to Facebook's Libra, according to a report by Bloomberg. He suggests a Libra-like currency should become the world's reserve currency.

At the U.S. Federal Reserve's annual meeting, Carney warned that while in the short-term central bankers have to deal with the situation; dramatic steps need to be taken. Specifically, Carney identified the dollar's status as the world's reserve as something that needs to be replaced by a form of digital currency similar to the Libra, believing that would be a better option than letting another fiat currency like the renminbi (China's reserve currency) replace the dollar.

Some Other Major Adopters

August 31, 2019: Cielo, a Brazilian payments giant, is looking to unleash cryptocurrencies onto the nation of over two hundred million people.

August 29, 2019: Cryptocurrency trading and payments in Portugal are tax-free, the European country's tax authority has clarified.

August 29, 2019: A United Nations official has said that cryptocurrencies make global efforts to fight cybercrime "significantly" harder, due to their pseudonymous nature.

August 28, 2019: A Coinbase report finds that fifty-six percent of the top fifty universities in the world now offer at least one course on cryptocurrency or Blockchain, a twenty-five percent increase from 2018.

August 28, 2019: Companies and governments are mapping out the Blockchain ecosystem in Africa.

August 28, 2019: The United States Air Force (USAF), the aerial and space warfare service branch of the U.S. Armed Forces, has signed a deal with Blockchain start-up Constellation Network to secure defense data.

October, 2019: It was reported that Bitcoin transactions played a role in the takedown of the world's largest dark web child sex abuse website, including the arrest of 300+individuals in more than thirty-eight countries. These arrests, and the shut down of the dark web website, were made possible because law enforcement traced Bitcoin payments between individual visitors of the website and the website's operators.

While the exact details of the child sex abuse ring bust are unclear, we do know that law enforcement organizations across the world were able to identify hundreds of individuals to arrest.

Approximately 15.8% of the world's population, or 1.2 billion people, live on the continent of Africa. Many of the promises heralded by Blockchain technology have been around financial inclusion, so naturally the continent has been chosen by many companies and projects in the sector.

Currently France's richest man and the world's fourth-richest man will be engaged in trading of cryptocurrencies such as Bitcoin (BTC), Bitcoin cash (BCH), Ethereum (ETH), XRP, Litecoin (LTC), and Dogecoin (DOGE).

Bernard Arnault, who owns brands like Christian Dior, has announced himself as pro-Blockchain.

Constellation Network Inc., a San Francisco based technology further said that it will provide "decentralized security" to USAF data pipelines with encryption, as well as create "audit trails and a live overview of the status of any data source." This is important because of the recent attacks on centralized systems like Equifax and Visa.

The USAF also partnered with SIMBA Chain to secure its entire life-cycle supply chain using Blockchain technology.

Bank of America—the second-largest in the US in terms of total assets—said that widespread adoption of cryptocurrencies and other fintech innovations could require the bank to make "substantial expenditures" to update its existing services and remain competitive with upstart firms.

Chief operations and technology officer at Bank of America Catherine Bessant stated: "We have around fifteen patents, in the Blockchain or cryptocurrency space and working on many more.

Bank of America and Goldman Sachs are also collaborating with members of the R3-led Blockchain consortium that is made up of forty-two global banks connected together on an Ethereum Blockchain R3-managed distributed ledger. So, the banks do want to get involved and manage control.

Some of the banks involved were Barclays, BMO Financial Group, Credit Suisse, Commonwealth Bank of Australia, HSBC, Natixis, Royal Bank of Scotland, TD Bank, UBS, UniCredit, and Wells Fargo.

David Rutter, the CEO of R3, said in a statement, "Securing the backing of forty-two of the world's leading banks demonstrates the level of interest in our initiative."

August 19, 2019: Fidelity Charitable, the financial services giant's charity unit, is now accepting XRP cryptocurrency, the world's third-largest cryptocurrency, for donations.

August 14, 2019: Samsung quietly adds support for Bitcoin to its Blockchain Keystore. South Korean electronics giant Samsung has silently added support for Bitcoin to its Blockchain Keystore, an online decentralized applications store.

August 9, 2019: Germany's biggest insurance company, Allianz, has been developing a Blockchain-based digital token for payments, similar to investment banking giant JPMorgan's JPM Coin.

Payments giant Mastercard is to develop a Blockchain-powered cross-border payments platform in partnership with enterprise-focused Blockchain firm R3.

In an announcement in September 2019, Mastercard said the two firms have inked a deal to "develop and pilot" the payments solution. It will initially be aimed at connecting faster payments schemes and banks backed by Mastercard's clearing and settlement network.

The platform will be built on Corda Enterprise, the commercial version of the platform, as opposed to the open-source Corda Network, R3 told CoinDesk.

The partnership is planned to merge R3's expertise at developing Blockchain solutions with Mastercard's existing payment systems and network. Ultimately, the firms hope the new platform will help tackle industry issues such as costly payments processing, liquidity management and a paucity of standardization and connectivity between banks and domestic clearing systems.

Among the leaders are the giant financial institutions Bank of America (forty-three patents), Fidelity (fourteen grants), payments leader Mastercard (joint second with twenty-seven) and, by far the leading tech player, IBM (another sharing second with twenty-seven patents).

It isn't surprising that the biggest financial institutions dominate the list of the largest Blockchain patent owners, some of the world's largest patent-owning companies, which have realised that advances in areas like Blockchain and have a major impact on business.

Chief operations and technology officer at BOA, Catherine Bessant stated in an interview withCNBC in Davos Switzerland, about Blockchain technology.

"I think we have somewhere around 43 patents, most people would be surprised at Bank of America with patents in the Blockchain or cryptocurrency space. Reportedly, they're drafting another twenty patents.

How much currency is traded every day? 5.2 trillion around the globe. It is hard just to get past the billions!

Goldman Sachs may be the next big bank to become involved in the world of cryptocurrencies. The U.S. Patent and Trademark Office published a patent for the investment bank that was, entitled "Cryptographic currency for securities settlement."

9

CASH IS DISAPPEARING

Governments and Central Banks around the world have really been picking up the pace in the war on cash in the last couple of years. Various former Central Bank and government officials openly declared the need to ban cash to stop "criminals" and "terrorists". In 2016, the ECB (European Central Bank) declared that it is literally starting to banish larger denomination bank notes. All of these actions will only help raise awareness and acceptance of Bitcoin and cryptocurrency. While attempts to ban cash can only help Bitcoin adoption and awareness, a look at the real reason for the push to ban cash uncovers some alarming truths about the existing financial system.

Central Banks are in full blown panic mode. The real reason why they are trying to ban cash is because there is hardly any non-counterfeit physical cash in existence compared to what is real. This is true when we look at actual currency in circulation (bank notes and coins) relative to aggregate measures of digits in accounts such as money market funds, equities, and bonds. Analyzing this data also magnifies how transparent Bitcoin is versus the banks reserve monetary system.

On April 14, 1865 Abraham Lincoln signed legislation that would allow for the creation of the Secret Service—not to serve as presidential bodyguards, but to help suppress money counterfeiting, which at that time was estimated to account for between one-third and half of the American currency in circulation.

The Board of Governors of the Federal Reserve System reports on its own website what amounts the trillions of dollars in circulation. Physical currency only covers less than 12% of what the Fed states to be the most liquid asset in the current financial system.

This is even more critical for the Fed if customers liquidate and demand physical cash.

If participants in increasing numbers started to demand liquidation of various types of accounts along with the possession of physical cash, the Federal Reserve would have an extremely serious problem (as would all banks in countries around the globe). The Federal Reserve would face a nightmare scenario where the general population started to question why they are being restricted from withdrawing banknotes from their own bank accounts. It is almost as if an attempt to ban cash is a not so subtle nudge to at least explore the option of exchanging some fiat for Bitcoin.

Hence, the Federal Reserve needs to try to ban cash, so nobody has the ability to demand it. If all accounts existed only digitally, then a run on the system at least would not involve the demand for physical currency. We have already seen restrictions on the size of ATM withdrawals.

This is all very bullish news for Bitcoin, of course. Bitcoin is a way to exit the Federal Reserve System which governs essentially all retail banks in operation in the United States. Bitcoin allows owners to have complete control over their own money. Will governments ultimately succeed in banning cash?

There is no doubt that it is under serious consideration. The fact that the Federal Reserve needs to even consider this option tells us quite a bit about the health and viability of the existing financial system and the potential of Bitcoin to completely disrupt it. So little physical cash exists the Central Banks have no option but to banish it and blame it on "terrorists" and "criminals".

The Rise of Electronic Payment Methods

Not only are there credit and debit cards, but there are also bank transfers, direct deposit, and online payments. What's more, there are now an abundance of systems that enable you to transfer money instantaneously to individuals, either by email or text.

It's simply too convenient to make payments electronically, particularly with the internet, given the fact that merchants and vendors can now be hundreds or thousands of miles away. Electronic payments just travel better than cash.

The Fight Against Crime

The invisibility of cash is more effective for the criminal element. There are a couple of potential issues here. The first is that heavy use of cash has been commonly associated with crime, particularly drug-related crimes. The elimination of cash may seriously impair criminal activity.

The other issue is theft. Today, a lot of people are afraid to carry cash, particularly large amounts. In many countries, people have no choice but to physically carry it and protect it, making themselves obvious targets for theft. While debit and credit cards can be canceled and replaced if stolen, once cash disappears, it's gone forever.

There would be job eliminations in banks and credit unions. There's little doubt the elimination of cash would enable banks, credit unions, and other financial institutions to reduce staff. After all, it takes more people to handle the business of cash transactions.

Electronic payment methods, by contrast, are completely digital. Far fewer people are needed to manage the process.

A 2017 Report by ForexBonuses.org

The following list shows the top ten cashless countries. The US is in the middle of the list at number five.

- Sweden
- France
- Australia
- Russia
- Canada
- USA
- Germany
- UK
- China
- Japan

When the 2008 economic crisis hit, central banks feared that investors would flee digital wealth for the comfort of cash. Indeed, the financial system was close to collapse when depositors attempted to pull $500 billion out of money market funds.

When it happened, the world's central banks realized that their worst nightmare could become a reality: if a significant percentage of investors/depositors ever tried to convert their 'wealth' into cash, the whole system would implode.

As a result, almost all the Fed's actions since 2008 have been aimed at forcing investors away from cash and into risk assets. The most obvious move was cutting interest rates to 0.25%, so the return on cash became insignificant. In Greece, they limited the withdrawals each day and later raided the citizen's financial assets/savings to pay back the debt.

It's already happening in other countries. France has banned cash transactions over one thousand euros Spain has banned cash transactions over twenty-five hundred euros. Meanwhile, Uruguay has banned cash transactions over five thousand dollars.

The continuing removal of higher denomination bills from several world economies, and the creation of centralized, state cryptocurrencies by governments worldwide cannot be ignored. All in the name of safety, security, and financial inclusion.

In India, Prime Minister Narendra Modi shocked the predominantly cash-based society into a panic, stating that the five hundred and one thousand-rupee banknotes would be withdrawn from the financial system overnight. They would no longer be legal tender. This forced people to take their now worthless one thousand and five-hundred-rupee notes to banks within fifty days of the announcement, to exchange them for smaller denominations. Now, The Royal Bank of India is moving to ban all cryptocurrencies and set up their own state-approved, digital rupee.

This is all part of a crackdown on corruption and illegal cash holdings. "Black money and corruption are the biggest obstacles in eradicating poverty," Mr. Modi said.

The move was said to be designed to lock out money that is unaccounted for—known as "black money"—which may have been acquired corruptly, or is being withheld from the tax authorities.

There will lots of questions of those who turn up with hundreds of thousands or millions in currency. Maybe to solve counterfeiting we emulate the Chinese in their transition from coins to paper money. The emperor had a good handle on both money supply and various denominations. In the place of where the American bills say, "In God We Trust," the Chinese inscription warned, "All counterfeiters will be decapitated."

Australia: The supposed coordination of governments and tech companies to create a one-world, cashless society is often viewed as a conspiracy theory.

Australia's proposed cash transaction ban for 2020, the continuing removal of higher denomination bills from several world economies, and the creation of centralized, state cryptocurrencies by governments worldwide cannot be ignored.

Australia's "Black Economy Taskforce" wants to put people accepting over ten thousand AUD in cash in the slammer for up to two years, or fine them up to twenty-five thousand, in a bid to fight black market economies.

New Zealand: New Zealand Reserve Bank governor Adrian Orr wants to tax cash holdings, simple as that, so people can't hold large lumps of physical cash.

Digital currencies are extremely convenient. Everybody in the world really should have a chance at financial inclusion. Settlements and payments can be made effortlessly with just a quick scan or entering a PIN, so it's no big deal.

Finland: In Finland, passengers on state railways won't be able to purchase tickets with cash for long-distance trips.

But this is not a perfect world. Governments are corrupt. Artificial monopolies and seas of red tape exist, keeping the life-threateningly impoverished and entrepreneurial from accessing crypto and banking services via strict identity issues.

A forty-dollar smartphone and an internet connection enable anyone, anywhere in the world, to make or receive money with Bitcoin. In the name of regulation, safety, and financial inclusion, however, the state makes the situation more chaotic, less safe, and extremely exclusive where real human need is concerned.

It a trend that is expected to move quickly to a slow position and stall in certain industries.

Sweden: Why are Swedes barely using cash?

Sweden is a unique economy when it comes to payments. It's home to a popular instant payment app called Swish, set up in 2012 by seven of the largest banks in the country. More than half of Swedish consumers are signed up to the app.

Ask most people in Sweden how often they pay with cash, and the answer is "almost never." A fifth of Swedes, in a country of ten million people, do not use automated teller machines anymore. Sweden's cash supply is being squeezed out so quickly that in five years—by 2025—they will stop accepting bills altogether. The concerns of going cash-free for any society are many. The head of Central Bank of Sweden states there are concerns, as they need to figure out what cash-free means for the young and old consumers. The Swedes lead the way for a cashless society. A fifth of the population does not use ATM machines anymore. If cash disappears, banks will have greater control.

In Europe, one in five say they rarely carry money. The majority of purchases are made with a payment system called Swish, set up by Sweden's biggest banks. About half of Sweden's 1400 bank branches refuse to take cash. Banks take on lucrative fees when people pay with debit or credit card and their Swish app. Bank robberies are way down, as only two were robbed in 2018, compared to 210 in 2008.

Their central bank plans to create their own digital currency in 2020 called the e-krona.

These trends signal a global push to kill paper money in the name of safety, security, and financial inclusion.

IOT- Internet of Things/Dark Web

If an Internet of Things is the series of protocols through which networked objects talk to each other, what is the equivalent for networked defense systems? What defines an Internet of Wars? While trying to envision this network space, technically it already exists. The U.S. Department of Defense has its own networks, and its own network of networks. In trying to understand what an Internet of Wars actually looks like and how it shapes the internet, you can't do much better than the history and constituent parts of the Global Information Grid (GIG), the Department of Defense's system that connects all of the military's networked systems to each other.

SIPRNet (Secure Internet Protocol Router Network), network was briefly in the news following a US Army personnel named Chelsea Manning released of thousands of secret documents (downloaded from secure government networks, via SIPRNet) to Wikileaks.

One Billion Websites

The deep web is the part of the internet that is not indexed by google or other search engines. A part of the deep web is the dark web, which exists inside layered proxy networks, so-called Darknets. Of these Darknets, Tor is by far the largest.

Tor is a browser that you can download on to your computer in order for you to access the Darknet. Tor is also well known to those that have been hit with ransomware. In order to pay the bad guys in Bitcoin they need to do it through the Tor browser.

But what kind of sites are hosted in this dark net? What do they do, and why are they there?

Why would one of the largest organizations known for its invasive stance on privacy have a Dark Web address? While Facebook might collect everything you say and do on its platform, it isn't happy with sharing this information with others. Facebook is also keenly aware of attempts by many governments to restrict access to a tool that allows strangers across the web to talk and collaborate freely.

Facebook has a DarkWeb address and it doesn't make it much easier to maintain an anonymous account, but it does make Facebook more accessible in places where it's censored.

Contrary to popular opinion, it's actually quite easy to link Bitcoin transactions together in order to identify if you know how to do it. This is because public Blockchains are totally transparent and browsable by anyone.

Still, dumb criminals are constantly caught for using Bitcoin in illicit activities. This is because Bitcoin is not anonymous. In fact, there are barely any cryptocurrencies on today's market that are capable of masking identities when sending, receiving, and spending cryptocurrency. So, ever wonder how these 'cyber-baddies' are turning ill-gotten money, too sketchy for use in the real world, into clean cryptocurrency? Bitcoin is easily laundered through unregulated exchanges. This is done by simply trading the Bitcoin a number of times across various markets. For example, a user can deposit onto an unregulated exchange, swapping it for various altcoins.

Each time a trader exchanges cryptocurrency for another, they are adding degrees of privacy similar to 'hopping' between wallet addresses. Although, how effective this is depends heavily on the exchange's monitoring technology, so this might not be a totally airtight solution.

The user can then withdraw their cryptocurrency to an external cryptocurrency wallet via other anonymous exchange accounts they own. Depending on the exchange, they could convert it to allegedly 'clean' fiat, but fiat markets on unregulated exchanges are hard to come by, and often short-lived.

Governments around the world along with large corporations set up security networks like honeypots. A honeypot is a decoy computer system for trapping hackers or tracking unconventional or new hacking methods.Banks cybersecurity teams are trying to make more informed decisions about threats with help from researchers who go undercover online to find intelligence on pending attacks or customer information for sale.

These researchers-for-hire can navigate the Dark Web in ways corporate cybersecurity teams can't.

They have expertise in infiltrating forums and have different techniques for getting credibility in those forums," said Jason Witty, chief information security officer at U.S. Bancorp. "Most companies wouldn't use such techniques," he added.

"U.S. Bancorp, Wells Fargo & Co. and other banks use vendors to gain an insight into attackers' activities, ferreting out information about distributed denial-of-service attacks and new malware.Corporate security teams integrate this data with threat information from other sources, such as the U.S. government and the Financial Services Information Sharing and Analysis Center," said Rich Baich, CISO at Wells Fargo.

"Typically, the security team would activate defenses after a major attack was underway," he continued. "With the Dark Web intelligence, the situation played out differently. We were able to do that before the attack started." he said.

10

MANAGING CRYPTO WALLETS

What's in your wallet?

Let's see: driver's licence, health and credit cards, cash, all of your (PII) personal identifiable information, and more. You also log onto the internet and do web searches with Google Chrome, IE, Firefox or other browsers. For email, using encryption will help keep your data secure. ProtonMail uses client-side encryption to protect email contents and user data before they are sent to ProtonMail servers, unlike other common email providers such as Gmail and Outlook.com. A few of you may use something like VPN Express. This keeps your online browsing private. Without a VPN, third parties can see your internet traffic. Your internet service provider can monitor your activity and share it with other organizations. Governments can use your data to restrict your access to content, and corporations can exercise price discrimination against you. And you're vulnerable to cybercrime and snooping, especially on public Wi-Fi.

Using public wifi can create a major security vulnerability—by placing a passive receiver called a wireless sniffer, in the vicinity of the wireless transmitter, that receiver can obtain a copy of every packet that is transmitted! These packets from you can contain all kinds of sensitive information, including passwords, social security numbers, trade secrets, and private personal messages.

With cryptocurrency, you are in effect your own bank with greater security!

All Bitcoins and cryptocurrencies need to be stored in your digital wallet and protected with your private key. Period. You have complete control, not the banks. This will be an uncomfortable step forward for all of us. You will be your own bank.

In setting this up, it may be difficult to understand the good guys from the bad guys. If you fall victim to a scam there is no recourse. Most of the scams revolve around an attempt to steal your coins/your private keys.

Websites that look like exchanges or web wallets but are fakes have plagued some of the top Bitcoin websites.

Stealing log-in information and misleading holders into sending Bitcoins is commonplace. “Send me your Bitcoin and I will ship the goods.” The bad guys will try to win your trust by sending fake id’s, profiles or impersonating other people you may know.

A crypto or Bitcoin wallet is simply an app, exchange website, or device that manages the private keys.

A “wallet” is basically the equivalent of a bank account. It allows you to receive Bitcoins and other coins, store them, then send or receive them. It holds your private key(s).

You can think of a wallet as your personal interface to the Bitcoin network, similar to how your online bank account is an interface to the regular monetary system to pay bills, send or receive money.

Why not use a trusted exchange? Aren’t they just like a bank that will protect me?With crypto, using a third party called an exchange for storage is just like giving the combination for your bank account. Well, you could use them if you only have a minimal amount of that currency while you educate yourself about types of wallets and proper storage.

With an exchange, you can buy and sell your crypto. However, if you store your crypto, you are leaving your private keys—secret codes that allow you to spend your coins—in their hands. In reality, it’s not coins that need to be stored and secured, but the private keys that give you access to them. Exchanges get hacked, lose track, or go bankrupt, and with that your crypto is gone. There is no recourse. Your private key is much more than a regular bank pin number and much more important.

Hardware Wallets

The Ledger Nano S is one example of a hardware wallet.

A hardware wallet is just like a USB key with built in technology, built for the sole purpose of securing crypto coins.

The two most popular and best Bitcoin and cryptocurrency hardware wallets are:

- Ledger Nano X
- TREZOR T

Hardware wallets are a good choice if you're serious about security and convenience, reliable Bitcoins, and crypto storage. Hardware wallets keep private keys separate from vulnerable, internet-connected devices.

Your all-important private keys are maintained in a secure, offline environment on the hardware wallet, fully protected, even should the device be plugged into a malware-infected computer.

If you use your computer, as Bitcoins and cryptocurrencies are digital, cyber-criminals could potentially target your computer's "software wallet" and coins by accessing your private key.

Generating and storing private keys offline using a hardware wallet ensures that hackers have no way to reach your coins. Hackers would have to steal the hardware wallet itself, but even then, it can be protected with a PIN code.

Don't worry about your hardware wallet getting stolen, lost, or damaged either; so long as you create a secret backup code, you can always retrieve your coins.

Which Wallet is Best for You?

Why are you using Bitcoin & Crypto? Investing or saving? Then, a hardware wallet will keep your coins safe.

Otherwise, a software wallet will send and receive Bitcoins and crypto just fine. Best of all, software wallets are free. Each wallet has pros and cons, and different wallets are built to solve different problems. Some wallets are geared towards security or privacy.

Each day, new Bitcoin scam wallets that are designed to steal peoples' Bitcoins are added to the Google Play Store and Apple app store.

The three most popular hardware wallets are KeepKey, Ledger Nano S, and Trezor.

They're expensive, but the price can be worth it if you own a significant amount of coins or, in the case of Bitcoin, a portion of a coin. A hardware wallet will protect a few hundred dollars worth of Bitcoin/crypto just as effectively as a few million.

How Hardware Wallets Work

Fast tracking, accelerated start-up companies, and others trying to adopt to the new technology will pale compared to the lightspeed of the adoption to digital currency. Thousands of currencies are already here, and thousands more will appear. The foundation for every new innovation is and will be THE BLOCKCHAIN. No digital currency will survive without the Blockchain technology. Understand the impact of THE NEW MONEY!

Since the investment goes into the cryptocurrency itself, you should ask: how do cryptocurrencies make money? This is a new field, and there is a lot to learn still. Cryptocurrency promoters argue the case as follows. Blockchain start-ups aim to create an "ecosystem": an economic system in which many participants come together to do business for mutual benefit.

Examples of ecosystems include marketplaces (e.g. stock exchanges), information exchanges (e.g. credit bureaus), communications platforms (e.g., email or Facebook), and payment mechanisms (e.g. Fedwire or Bitcoin). The value of these ecosystems increases rapidly with the number of participants (i.e. "network effect").

As a result, in a given domain the ecosystem that grows fastest can become a hugely valuable and powerful natural monopoly (e.g. Facebook). By design, the cryptocurrency is required to access the ecosystem: if you want to do business in the marketplace or trade your information, you have to buy and sell using the cryptocurrency.

Hence, the success of the ecosystem will create demand for the cryptocurrency: the greater the number of participants and transaction quantity and value, the more people will want to buy and hold the currency. And, the entrepreneurs usually pledge that the supply of the currency will be fixed or severely limited. So, increasing demand should drive up the price of the currency. Bitcoin itself is a simple example. It was designed primarily as a medium of exchange (a way to pay) that is secure, global, easy to use, and free from control by banks or governments. People buy and hold Bitcoins to be able to access this service conveniently, which creates demand for Bitcoins. Growth in the volume and value of Bitcoin transactions adds to this demand.

According to Dan Boneh, co-director of the Stanford Computer Security Lab and professor of cryptography, he is attracting a large number of people to his cryptocurrency courses. As a cryptography researcher for almost thirty years, Boneh began teaching a Bitcoin and cryptocurrency class in 2015, attracting more than one hundred students. His online cryptography course has seen over one million signing up.

In the third quarter of the 2019, Freelancer listed 480,000 new jobs. To date it has more than twenty-five million active users on its site.

You are responsible for the security of your Bitcoins and that's why you keep it in a wallet that you have 100% control over. This is done by having the ownership of seed word or private key.

To simplify your understanding of Bitcoin wallet security, you just need to use a well-recognized wallet that lets you, and only you, keep the seed words. These seed words are the password for your Bitcoin.

Even if you lose your phone or hardware wallet, you can recover your Bitcoin using the seed words.

There are different kinds of wallets for Bitcoins:

- Mobile wallets which come with a mobile application and are available for both iOS and Android.

- The desktop wallet which comes as software.
- Web wallet which comes as a website.
- Hardware wallet which is the most secure.
- A paper wallet, which is also secure, but difficult to set up .

The challenge with the mobile app-based wallet is the security that is partly dependent on the security of mobile OS that makes it not so secure. The same is true of the desktop-based wallet. The most secure way to keep your Bitcoin is in a hardware wallet.

Only buy from an exchange that is trusted and recommended by reliable resources.
Treat your Bitcoin investment like any other investment.

When selecting a reliable Bitcoin wallet, irrespective of whether it is a software or a hardware wallet, one should judge it based on the following criteria:

- Control private keys: A wallet where you own and control your keys.
- Backup and security features: Seed backup keys and pin codes.
- Hot/Cold Wallet: Whether a wallet is a hot or cold wallet.
- HD Wallet: A wallet that generates new addresses itself.

One has to be serious while choosing a Bitcoin wallet, because the wallet is something where your private keys and public keys reside, which if compromised can result in you losing all your Bitcoins.

The purpose for which you are going to use a particular wallet also plays an important role in selecting which type of wallet is required of you. Some of the purposes are:

- Long-Term Savings
- Short-Term HODL
- Daily Transactions
- Daily Buy/Sell or Trading of Crypto

Survey Bitcoin Reddit or the Bitcoin Forum to find out as many details about new wallets before purchasing.

"The private key must remain secret at all times because revealing it to third parties is equivalent to giving them control over the Bitcoins secured by that key. Forever lost, too." — Andreas M. Antonopoulos, Mastering Bitcoin: Unlocking Digital Cryptocurrencies

The US dollar is already a digital currency. It runs on a Blockchain with one master node called The Federal Reserve. It has no supply cap, can be reversed, created, or destroyed at will.

Most of what people thought about money is actually credit. The total amount of credit in the US is over fifty trillion dollars, but the actual money in the US is only about four trillion.

Rising prices are only a symptom of an expanding currency supply. If more credit is given, more goods and services are purchased. Even if a person or company's income is growing, it is not enough.

September 19, 2019: The Federal Reserve jumped into financial markets for a third straight day on Thursday in another attempt to keep interest rates from moving higher. The central bank has injected a total of $203 billion into markets this week.

The Federal Reserve is a private corporation and not government. It is the Central Bank of the USA and owned by banks around the world. The monetary system is very complex—not what is out in front of us and how we use it, but how it actually works.

There is a lot of misunderstanding about the difference between currency and money. Currency is a medium of exchange. A unit of account. It is portable, durable, divisible, and something called fungible, which means each unit is the same. Each dollar in my pocket is the same as a dollar in your pocket. They're interchangeable.

Money is all of the above, plus a store of value over a long period of time. The currency in your pocket is a medium exchange because it has numbers on it. It's divisible because you can make change.

But because the government can print more and more of it, it dilutes the currency supply and actually transfers wealth away from the people, to the government, and into the banking system.

Gold and silver are the optimum form of money because of their properties. A unit of account is the same all over the planet. The government cannot print commodities.

Over the centuries, there have been over a thousand currencies that were not backed by gold or silver, and 100% of them failed. The currency that is still here today, from five thousand years ago is gold and silver.

Every country in the world uses Fiat currency (cash), which is backed by nothing.

Money must be a store of value and maintain its purchasing power over long periods of time. In the last ninety-five years, the dollar has lost 90% of its purchasing power. Gold and silver are the ultimate store of value.

The government continues to print more currency and adding to inflation. The food has not changed; the real estate has not changed. Prices for food and real estate are a reflection of the currency going down, not prices going up.

11

COW DISRUPTION

One of the biggest reductions to the carbon footprint in animal agriculture might happen soon.

Here we are again with disruptive and exponential technology.

There are dozens of companies—more than just start-ups—around the globe that are on the brink of creating the exact look, feel, and taste of meat. NASA has even created meat in space. There is already market penetration with a company such as Beyond Meat, and the largest fast food giants already have replacement burgers on the menu. Good news for all the vegans, but what more is at stake here?

The US Food & Agriculture report about food consumption around the globe is mind blowing; every ten seconds 24k livestock animals are killed. There are seventy-five billion killed each year. There are one billion pigs, one billion sheep, and twenty-three billion chickens on the planet. The overall demand for meat is accelerating significantly as emerging economies slowly moving up the economic scale are starting to eat meat or more of it. The additional land mass required for that growth is unsustainable. Massive energy is needed, deforestation is increasing and now more than 70% of the antibiotics produced in the US go to animals.

What if we could slow this trend? It would reduce the carbon footprint. What if we could significantly accelerate the speed of current studies that are happening right now in the realm of molecular innovation?

The key to understanding the speed and scale of this disruption is when the most essential contents of a product are replaced quickly and cheaply. In this case it is animal protein. This is a protein disruption driven by economics we now recognize are alterable.We can now program microorganisms to produce almost any protein we want because of huge advances in precision biology (the convergence of biology and information technologies) via a process called precision fermentation.

What role does Blockchain play here?

The majority of scientists in all fields withhold data because it cost too much to send the materials to the scientists who had requested them, which slows the process of publishing their findings. When data is a physical thing such as an antibody, a chemical, a mouse, or a reengineered organism, the cost and administrative difficulties are an important obstacle. Also, the lack of a centralized digital storage space means that data might only be kept on a personal computer or exist solely in paper form, especially for scientists who have hundreds of studies under their belts.

What if there was a forum where scientists, engineers, people in genetics, biology, and chemistry, could access all of their studies on a Blockchain—a private Blockchain that was open source for all the scientists to see everyone's studies and results. It would unleash overwhelming euphoria amongst them. But, more, it would create exponential technology in agriculture to DNA.

All of this may be leveraged with openness transparency, competition, and trust—The Blockchain solution.

Another significant component is what is behind the amount of energy and water needed for livestock production today and in the future. That is the global crisis for a supply of fresh water. It is not so much a lack of fresh water but how it is used. Blockchain and cryptocurrencies can be used as an incentive to reduce and use water wisely may have an impact.

In 2018, Cape Town South Africa had extreme water shortages and finally declared a Day Zero where water would be shut off and rationed out. There was a drastic reduction and Day Zero has been pushed off. Ironically, ¼ of the water supply is used in the South African wine industry. Mexico City once had a lake surrounding the city but over the centuries that has been drained off to make room for twenty-two million people. Mexico City is tapping into the underground supply of water, causing the city to start sinking. In many parts of the world, such as California, water is extensively used in some of the most arid land areas to grow Alfalfa for cattle food. India and China both grow their most intensive crops in their driest areas.

A major part of the freshwater supply goes into agriculture. Going through the lifecycle of a cow, the consumption of water used is approximately 1600 litres or 400 gallons of water to produce just ¼ pound of beef. Farming livestock represents up to 18% of global emissions. Blockchain it appears will have a significant role in the advanced technology.

Food as a software is about to change the food industry. Engineered by scientists at a molecular level and then uploaded to Blockchain databases. Food designers around the world will work like software developers. Breaking down a cow into meat parts can be replicated molecularly.. Scientists working together on a global scale using the new technology would fast forward new ideas and solutions.

Disrupting the Cow/Carbon Footprint Reduction/ Climate Change

By 2030, scientists believe and estimate the number of cows in the U.S. will have fallen by 50% and that the beef and dairy industries will be collapsing, as animal-derived foods are replaced by modern equivalents.These industries will still be here twenty years from now, but expect them to look entirely different. But what will the farmers do? They will need to change their product line.

A fast-moving, improving technology business model's fast falling costs created the deepest, most consequential disruption of food and agriculture in the last ten thousand years; we face the end of the cattle industry as we know it. Well, there may be a major impact in the reverse trends by 2030, but decades before we see animal beef off the menu.

If the U.S. embraces the cheaper and more nutritional modern foods, it is poised to create one million jobs and usher in growth to one trillion annually by 2035. If it resets its risks, locking in expensive and obsolete assets technologies and skillsets, other countries will capture the jobs and wealth that comes with building a world-leading industry.

The key to understanding the speed and scale of this disruption is to recognize what happens when the most essential contents of a product are replaced quickly and cheaply. A bottle of milk contains only 3.3 percent protein, so replace that and there goes the need for a dairy cow's protein. In Silicon Valley, we expect these newfangled proteins to reach cost parity with animal proteins in the next three to five years.

All other livestock industries will suffer a similar fate, from the supply chain to processing plants to crop farmers to tractor manufacturers; they all will be hit hard.

Today 90% of American-made cheese uses Precision Fermentation protein. This is not genetic modification of foods. The cost of PF is falling exponentially from a one-hundred thousand dollars per kilogram to a mere two-hundred dollars. The cost will continue to go down.

Further, PF proteins will be five times cheaper than animal proteins by 2030 and ten times cheaper by 2035. Mainstream analysts fail to see the speed scale and impact of disruption because their models do not account for the complex, systematic ways disruption models unfold.

This is not a simple substitution. As the protein drive from animals is replaced by less expensive, higher-quality PF alternatives, it becomes more expensive to produce animal products in the face of decreasing demand, triggering a death spiral, increasing prices in reversing economies of scale. The ultimate decision will be with us - the consumer. There are already 100's of companies who have meat replacements in our food stores and now fast food. If it looks like, taste like, has equal to or better vitamin or protein amount - and then the cost is 75% less, well how would you choose?

Well, this disruption is inevitable. Decision makers can affect how fast it happens, how wide it spreads, and who benefits. This market will be designed to promote openness, transparency, and competition.

Ten years ago, the Bitcoin Blockchain launched, with the intention of becoming the future of money. Today:

- Goldman Sachs, one of the world's biggest banks, has built its own Blockchain.
- The Australian government has signed a billion-dollar deal with IBM to bring Blockchain to the public sector.
- The United Nations has launched an initiative to improve humanitarian efforts using Blockchain technology.
- The city of Dubai announces it will be the world's first Blockchain-powered government by the end of 2020.
- IBM starts offering BaaS (Blockchain-as-a-service) to businesses.
- Zcash, a new cryptocurrency with unprecedented security, is launched.
- Google joins IBM, Amazon and Microsoft in testing Blockchain services with clients.

In 1999, Nobel-Prize-Winning economist Milton Friedman said, "The one thing that's missing is a reliable e-cash, whereby on the internet you can transfer funds from A to B without A knowing B or B knowing A," and that "it would be one of the major forces for reducing the role of government."

Since Bitcoin's code is open source (available to the public), people can start making their own coins. Namecoin, Litecoin and Swiftcoin, appear on the scene. All have their sights set on becoming a global currency.Vitalikb Buterin, a nineteen-year-old developer who co-founded Bitcoin Magazine, has a vision for a new kind of cryptocurrency. One that can do more than simply make payments. He originally proposed this in his whitepaper called 'Ethereum: A Next-Generation Smart Contract and Decentralized Application Platform', which now has the largest crypto market cap behind Bitcoin.

2016

Refugees Building Blocks is the largest deployment to date of blockchain technology for humanitarian purposes.

The WFP (World Food Programme) successfully runs an experiment to serve 100,000 Syrian refugees using blockchain technology.

WFP has been rolling out blockchain technology, using the distributed ledger technology to expand refugees' choices in how they access and spend their cash assistance. WFP is also exploring whether Building Blocks can make cash transfers more efficient, secure, and transparent. Most notably, WFP has been using blockchain to deliver food assistance more effectively to 106,000 Syrian refugees in Jordan. Building Blocks facilitates cash transfers while protecting beneficiary data, controlling financial risks, and allowing for greater collaboration.

Parity Technologies, a company that enables businesses and or ganisations to capitalise on blockchain technology, worked with the WFP to create a private blockchain network for processing cash transfers. Part of the UN WFP' project is the private blockchain is built using cryptocurrency Ethereum.

The blockchain allows the WFP to account for balances without the need for a third-party financial service provider to act as an intermediary, eventually, this will reduce bank fees and other administrative costs. It started as a pilot project with ten thousand users in 2017 and has grown more than ten-fold in the last year. With the help of the Building Blocks blockchain, the WFP has been distributing aid more securely and efficiently to Syrian refugees in Azraq Refugee camp in Jordan.

The Journey So Far

In the last year, the Building Blocks project had the following impact:

1. More than 100,000 refugees have been served on a monthly basis.

2. The UN saved more than 40,000 USD per month in bank transfer costs.
3. Tens of millions of USD of aid has been delivered through the blockchain.

In addition, the system has helped create blockchain-based immutable identities for refugees. Having a permanent digital identity helps refugees apply for and receive other aid and facilities.

Here's how both the UN agencies will collaborate on using the Building Blocks network to help women and men receive aid:

A woman for example, will scan her eye to request cash back at WFP-contracted supermarkets. This will link to her account on the blockchain, and the amount of the cash distribution is automatically sent to Building Blocks. The fact that UN Women and WFP validate each other's transaction through a common blockchain network, results in improved security and accountability. There are also opportunities for cost and risk reduction, as well as increased harmonization of aid efforts.

The WFP's approach was to launch the Innovation Acceleration Program in 2017. The accelerator program has funded eight projects, including a low-cost hydroponic growing system for desert use, fintech apps for farmers in Zambia, and experiments involving artificial intelligence, drones, and satellite imagery for disaster recovery efforts.

The program integrates blockchain with biometric technology to create digital profiles unique to each family. The residents don't have to rely on having cash on hand or a physical debit card to go shopping. Instead, to buy food and supplies, camp residents stare into an iris scanner at the register to verify their identity. In the blink of an eye, the scanner cross-checks their identity against a UN database, queries their account, and allows them to purchase what they need. At the end of the transaction, they receive a text message via their cellphone informing them of their balance, plus a receipt of what they've just spent.

It is fast and secure. It streamlines the system. It avoids corruption. It eliminates cash needs or to have to hold onto cash in order to protect against theft.

Blockchain advocates envision a day when refugees can re-establish their identities through bioinformatics, a digital ID, and financial documents linked to a Blockchain-based system, which they own.

For example, the Finnish Immigration Service has partnered with a Helsinki-based Blockchain company called MONI. In 2015, the Nordic nation began issuing refugees a prepaid Mastercard that's linked to a digital identity number stored on a Blockchain ledger. Even without the documents needed to open a bank account, MONI cardholders can receive benefits from the government and use the cards at Mastercard terminals to conduct basic financial transactions. MONI handles the secret cryptographic handshake required to process billing and payments.

Meanwhile, Accenture, Microsoft, and other companies have joined forces with the UN to form the public-private consortium ID 2020. The aim is to provide digital identities to 1.1 billion people who lack proof of their existence, in a manner that answers the basic questions of who they are and where they come from. Much like Building Blocks, the initiative will combine biometric data and the Blockchain ledger to create an encrypted and portable ID for refugees. If the effort is successful, refugees equipped with smartphones will have access to their personal data at all times.

As the World Bank points out, smartphones with built-in biometric sensors are becoming less expensive with each passing year. Between 2010 and 2017, the overall price of smartphones has declined by twenty-seven percent. By 2021, analysts predict 4.8 billion biometrically enabled devices will be in circulation. As their numbers increase, mobile ID schemes are expected to gain traction.

Building Blocks is looking to increase its presence within the aid sector. It remains to be seen if other agencies will adopt Building Blocks, or whether the donor nations that keep the WFP afloat will accept the Blockchain ledger as a faster and better way of doing business.

Nevertheless, more than one hundred Fortune 500 companies, cryptographers, academics and veteran Ethereum developers have joined in the Enterprise Ethereum Alliance (EEA) to develop roadmaps for enterprise and industry-wide applications on the Blockchain. Even in 2016, the World Economic Forum said, more than ninety central banks were holding discussions on Blockchain systems. Some eighty percent of banks were predicted to have started projects by the end of 2017 and in 2020 they are very close to the target.

Sweden is already using the Blockchain to record its property transactions. Estonia uses the system for registering its inhabitants and carrying out many business transactions.

The United Nations will use Blockchain to solve global social problems. WEF observed: "The global estimates for children under the age of fourteen [without identify documents] exceed six hundred million children worldwide. These children are literally 'invisible' to governments or development agencies that design and deliver social programs."

- Camera company Kodak announces plans to create its own Kodak Coin. Shares rise 60%.
- Switzerland begins accepting taxes in Bitcoin.
- America's biggest retail store, Walmart, runs a trial with IBM to track and verify the source of food on the Blockchain.
- Cryptocurrency EOS raises $4bn in the biggest project ever. Many believe it will one day replace Ethereum.

Car Leasing & Sales

The experience of leasing, buying, or selling a vehicle is a notoriously fragmented process for stakeholders on all sides of a transaction, but the Blockchain could change that. In 2015, Visa partnered with transaction management start-up DocuSign on a proof-of-concept project that used Blockchain to streamline car leasing—transforming it into a "click, sign, and drive" process.

Insurance companies will experience a major overhaul in all transactions. Facilitating claims management for property and casualty (P&C)insurers on DLT can automate processing through smart contracts, improve assessment through historical claims information, and reduce potential for fraudulent claims.

Impact to Insurance Institutions:

- Smart contracts can automate claims processing through third party data sources and codification of business rules.
- Distrubted Ledger Technology can drive reductions in operating costs through process simplification.
- Storing historical claims information on the ledger will enable insurers to identify suspicious behaviour and improve assessment.
- Building a comprehensive set of asset profiles and history will be possible.
- Adopting standards for relevant claims data will be necessary.
- Providing a legal and regulatory framework is the next step.

Key Ecosystems Stakeholders

Key Findings from the World Economic Summit

Distributed Ledger Technology (DLT) has great potential to drive simplicity and efficiency through the establishment of new financial services infrastructure and processes. DLT is not a solution for all problems, instead it should be viewed as one of many technologies that will form the foundation of next generation financial services infrastructure.

Applications of DLT will differ by use case, each leveraging the technology in different ways for a diverse range of benefits.

Digital Identity is a critical enabler to broaden applications to new verticals; Digital Fiat (legal tender), along with other emerging capabilities, has the ability to amplify benefits. The most impactful DLT applications will require deep collaboration between incumbents, innovators and regulators, adding complexity and delaying implementation. New financial services infrastructure built on DLT will redraw processes and call into question orthodoxies that are foundational to today's business models.

By shifting real power towards citizens and giving them real opportunities for prosperity and participation in society, maybe this could change the thinking of today's institutions. Or maybe we'll build new institutions. Fossil fuels are still widely used because of the low cost, comparative to renewable energy resources. So, can Blockchain help reverse that?

UN Supports Blockchain Technology for Climate Action

A major obstacle to addressing climate change is that our energy system is highly centralized, while the technologies we will need to cut carbon pollution, by and large, are not. As the price of solar panels continues to drop, more people will generate their own electricity from rooftop solar.

Rooftop solar owners want a way to sell their surplus power. Currently, in many places, they can sell that electricity back to the power utility, but this undercuts the utility's core business of generating and selling electricity. In parts of the country, utilities have gone to great lengths to curtail this practice.

Blockchain technology is expected to become an important means for effective control of carbon emissions, which is of great significance to China, the world's largest source of carbon emissions.

Experts say it could be used to track emissions, cut pollution and sell clean power. At every turn, it could make that work cheaper, faster and more transparent. 91% of business leaders believe that transparency builds trust. The same is true of consumers.

Powered by Blockchain and open data, our platform gathers and shares key product information and journeys in a way that's secure, trustworthy, and accessible. We envision a future where every physical product has a digital history, allowing you to trace and verify its origins, attributes and impact.

A Secure, Open Registry

Until now, centralized data systems were the only way to power a transparency and traceability system for materials to ensure data was trustworthy. Blockchain technology changes this fundamentally. We believe it can disrupt how we track the attributes and journey of every material thing, powering a system everyone in the supply chain can be part of.

Supply Chain Management

Most people prefer to buy products that are ethically produced. Products go through many hands before they reach consumers, and it is easy for producers to lie or distort what materials and chemicals they use, where they dump garbage, and how fairly they treat their employees. From the producer to the shelf: waste, inefficiency, and fraud are always possible.

Is the food in your pantry labelled? If so, can you even trust those labels? Your organic produce and meat may not be organic after all. The Ethereum based application, which will provide an online, fraud-free, food-tracking system from farm to store—giving confidence back to consumers that the produce/meat they are buying is truly what it says on the label.

All the way through the food chain, products will be tracked using both photographic and digital tracking data. This data will be uploaded to the Blockchain through each step of the food chain process. This data is then immutable and fraud-free.

Walmart

Walmart has been working with IBM for over a year, exploring uses of the blockchain to digitize the food supply chain process. In fact, the supply chain is one of the premiere business-use cases for block chain other than digital currency. Walmart is using the IBM Food Trust Solution, specifically developed for this use case.

Frank Yiannas, vice president of food safety for Walmart said in statement, "Each week, over 275 million customers and members visit our more than 11,300 stores under 58 banners in twenty-seven countries and eCommerce websites. Most supply chains are bogged down in manual processes. This makes it difficult and time consuming to track down an issue like E. coli in lettuce or beef. By placing a supply chain on the blockchain, it makes the process more traceable, transparent and fully digital. Each node on the blockchain could represent an entity that has handled the food on the way to the store, making it much easier and faster to see if one of the affected farms sold infected supply to a particular location with much greater precision."

"Before moving the process to the blockchain, it typically took approximately seven days to trace the source of food. With the blockchain, it's been reduced to 2.2 seconds. That substantially reduces the likelihood that infected food will reach the consumer.Suppliers don't have to be blockchain experts by any means. They simply have to know how to upload data to the blockchain application."

"Our customers deserve a more transparent supply chain. The present way for existing food traceability was outdated for the 21st century. This is a smart, technology-supported move that will greatly benefit our customers and transform the food system, benefitting all stakeholders."

Re-Cycle To Coin

Plasticbank.com uses Blockchain and empowers recycling ecosystems, incentivizing communities to recycle. The impact goes beyond ocean health: new economic opportunities are created for the world's most disadvantaged communities. Plastic Collection Credit is spent and kept in the local economies of our operating regions.

A recycling program on the Blockchain that provides a cryptographic token reward in exchange for depositing plastic cans or bottles. This means there is a whole new economic cycle that encourages people to collect plastic, and the people/community get rewarded.

Positive Blockchain.io is an open-source database, media platform and community exploring the potential of Blockchain technologies for social and environmental impact. We like to call them Blockchain Positivists.

Positive Blockchain projects are initiatives from start-ups, corporations, governments, or non-profit organizations, using Blockchain and decentralized technologies as tools to create a positive impact and solve social or environmental issues.

The true economic value of plastic is to make plastic too valuable to haphazardly throw away. In doing so, they created a system for plastic exchange, similar to any other currency exchange. They use an IBM-powered mobile app platform uses world-class banking and transaction technology to empower our Collectors to open a functioning digital bank account—sometimes, the first they've ever had.

Energy

Traditional power grids are centralized, which can create inefficiencies in energy distribution, like having unused surplus. And in parts of the world affected by natural disasters or poverty, power outages can leave people without access to electricity.

A peer to peer blockchain based energy system system would reduce the need to transmit electricity over long distances.

Changing incentives

In our world of complexity, it can be difficult for individuals or companies to see the direct effects of our actions.

Blockchain technology can help both individuals and companies to see the real impact of their actions and incentivize them to take the actions that benefit the environment.

The blockchain can be used to transparently track a variety of data like the carbon footprint of each product, the greenhouse gas or waste emissions of a factory, or a company's overall history of compliance to environmental standards. Companies and individuals can be incentivized to act in an environmentally sustainable way with the 'new money'.

Blockchain could discourage corporations and governments from back-pedalling on their environmental promises or misreporting their progress, because the technology would allow to transparently track important environmental data, and show whether commitments were met. Once the data is entered into the public blockchain, it stays there forever.

Non-profit Agencies

Personally I would donate more if I knew where my dollars were actually going to. If there was complete transparency with funds going to the actual cause or rather to 10% or 70% pay administration. Would everyone not consider to help out more? Blockchain is the solution.

12

MANAGE YOUR ENTRY TO BITCOIN

Is it time to put your life savings into Bitcoin? Ahh no, not really.

Lazlo didn't either but he did use 10k Bitcoins worth/$41 at that time to pay for two pizzas.

The (now famous) Bitcoin Pizza Guy, Lazlo Hanyecz, made the first real-world payment in Bitcoins (BTC). He paid ten thousand BTC for two Papa John's pizzas. With the pizzas worth tens of millions just a few years later, May 22nd is officially crowned Bitcoin Pizza Day.

You could buy a lottery ticket for the chance to win ten million, or you could have invested in these companies that you should have invested in 2014:

Stock returns since Satya Nadella became Microsoft CEO in 2014:

Microsoft: +320%; Apple: +233%; Facebook: +178%;
Cisco: +152%; Intel: +144%; Salesforce: +136%;
Google: +106%; eBay: +73%; Oracle: +60%;
Tesla: +36% IBM: +1%.

But if you really want to learn about the technology, then buy a small piece of Bitcoin—even$50-worth—and go through the process.

Buy it and hold it or just buy it and spend it. Either way, the knowledge gained will be immense no matter what life stage you are at right now. There are over 100k companies that accept Bitcoin and many websites to look at as well. Spendbitcoins.com is one site where you just search the product you are looking for. Online retailers were only the first wave. Several other large online retailers, like eBay and Amazon.com, followed suit. Other major companies to accept Bitcoin payments include Target, CVS, WordPress.com, Subway, Victoria Secret, PayPal, Expedia, Home Depot, Kmart, Sears, the Apple App Store, Grooveshark, Dell.

Ironically, many of those companies aren't exactly taking payments in Bitcoin. Instead, they are working with a middleman to immediately convert Bitcoin into U.S. dollars.

Even Overstock.com, the first major company to accept Bitcoin converts 90 percent of Bitcoin revenue into U.S. dollars.

Bitcoin Investment

With every passing week, the world's interest in cryptocurrencies is multiplying at a staggering rate. Whether it's Bitcoin, Ethereum, Litecoin, or any other cryptocurrency on the block, people around the world will own and trade cryptocurrencies.

At some point, financial planners, accountants, banks, or any financial advisory will need to be up-to-date in their advisory for you to invest or not. At this time, they will only recommend not to because it is off the charts, even for a high-risk, speculative investment.

One of the simplistic and easiest ways is to go through a Bitcoin ATM. You first research which mobile wallet you want to download. You need this for the ATM to scan your wallet address in order to send the Bitcoin. The ATM prints out a paper receipt also. Just like withdrawing cash from ATMs you follow the basic steps. Keypad the amount in cash you want to purchase and the ATM shows you the amount in Bitcoin. There are lots of youtube videos to break this down.

August 29, 2019: It's Now Possible to Lend and Borrow in Digital Currency. Here's How It Might Work.

So far, cryptocurrencies have failed to come anywhere close to widespread consumer adoption for payments.

Small progress is being made, however. For instance, it is now possible in August 2019 to borrow Bitcoins through a process known as “decentralized lending,” RBC’s(Royal Bank of Canada) Mitch Steves writes,

The learning curve on understanding the cryptocurrencies is going to take time. The risk factor is extremely high. The new money is highly speculative and volatile. The majority of buyers today are pure high-risk speculators buying and selling on a regular basis and are very comfortable doing so. Others are buying on the dips.

Is it better to wait until the world of crypto stabilizes before buying? What would be your reasons?

Only buy what you can afford to lose with any speculative purchase. Many are doing their entry points into the market buying a piece of Bitcoin or other and HODL (holding) while they start to increase their awareness and education.

Types of Exchanges

There are three categories of cryptocurrency exchanges. They’re as follows:

Brokers: Through these brokers, anyone can go and buy the desired cryptocurrency at a pre-set price. Most of these cryptocurrency brokers are similar in function as foreign currency exchanges.

Platforms for Trading: These platforms specialize in connecting buyers and sellers. For every transaction that takes place through the platform, they charge a fee—which is also how they make money.

Direct Trading: The way direct trading differs from the other two is that there’s no fixed price. The seller decides the amount, and the buyers have to determine if they want to buy or negotiate for any particular cryptocurrency. Direct trading is usually peer-to-peer.

What are the key attributes to look at while deciding to Trust a Blockchain exchange?

The meteoric rise of Bitcoin has also contributed to a significant increase in con-artists and exchanges which are mostly scams. So, before you decide to put your hard-earned money into cryptocurrencies, it is crucial to do your research. Below are the three things to know about an exchange before trusting them.

Reputation: In the age of the Internet, it's not too hard to find reputable reviews about anything—including Blockchain exchanges. Right from forums to commercially reviewed websites, you'll find loads of information that'll make it easier for you to decide whether an exchange is a right fit for you or not.

Fees: Before you even think about joining an exchange, do make sure you're well-aware of all forms of costs that the exchange charges. The fee-model should include comprehensive details regarding withdrawal, deposit, as well as transactional fees. Compare it with other exchanges and then make an informed decision.

Verification Systems: Any exchange worth its salt will have strong fortification in the form of user verifications and securities. While the exhaustiveness of the verification might seem annoying while registering, it'll protect your cryptocurrency and money from all sorts of scams.

Some Trustworthy Exchanges

Coinsquare: The first thing you should know about Coinsquare is that it's built on the same technology as the New York Stock Exchange. Yes, that's right. According to their website, they manage their ledger at least 2346 times in a single day to ensure security. Reputable and secured, Coinsquare is a popular choice for both crypto-veterans and beginners alike.

Coinbase: Not only is Coinbase backed by some of the world's best investors, but it's also trusted as an exchange by millions already. One of the key features of Coinbase is that its insurance covers the stored currency. So, there's that added layer of security. There's a digital wallet also available as a mobile platform through which users purchase Bitcoin, Ether, and even Litecoin.

Bitcoin Meetups/Endless YouTube Hustlers ICO's—The High-risk Game (IPO's)

In today's world, a corporation can offer shares to the public, allowing the company to raise capital from public investors purchasing the stock on the stock Exchange.

As a business grows, the founders having invested a chunk of their own money to raise capital as their business starts to gain traction, and the investment from the public allows them the opportunity for growth. The business has many valuations from industry analysts to help inform investors.

It must also comply with a strict set of reporting guidelines, as established by the Securities and Exchange Commission.

Initial Coin Offering (ICO)

The only similarity to an IPO (Initial Public Offering) when a company issues a new stock, is the purpose, which is to raise funds for new ventures. Purchases are made on independent exchanges geared for cryptocurrencies. They should be labeled as extremely high-risk. In fact, even higher because 95% don't make it. The majority turn out to be well-marketed and designed projects. Another group has real intentions of worthy projects but lack expertise. Project plans change and cannot sustain interest to create greater groundswell for continued investment to keep the project moving forward.

In an ICO, a project creates a certain amount of a digital tokens and sells them to the public, usually in exchange for other cryptocurrencies such as Bitcoin or Ether.

The public could be interested in the tokens on offer for either or both of the following reasons:

1) The token has an inherent benefit. It grants the holder access to a service, a say in an outcome, or a share in the project's earnings.

2) The benefit will be in increasing demand, which will push up the market price of the token.

Tokens, especially those of successful sales, are usually listed on exchanges, where initial buyers can sell their holdings and new buyers can come in at any time.

Crowdfunding is now used bringing together various individuals who commit money to projects and companies they want to support. It's a young and quickly growing market and it's transforming how people behave with their money. It's also transforming the ways businesses raise capital. As a type of digital crowdfunding, token sales enable start-ups not only to raise funds without giving up equity, but also to kickstart the project's adoption by incentivizing its use by token-holders.

Buyers can benefit from both the access to the service that the token confers, and from its success through appreciation of the token's price. These gains can be realized at any time (usually) by selling the tokens on an exchange. Or, buyers can show their increasing enthusiasm for the idea by purchasing more tokens in the market.

The number of ICO's I have seen in the last few years all have looked like worthy projects. Some were inspiring and appealing in terms of what they could possibly become. In a way, the excitement of the marketing and idea made it more like looking at the early start-ups Amazon and Spotify, knowing the challenges they may have but more what they could become.

The largest token sale in 2015 was Ethereum. Over 50 million ether tokens were created and sold to the public, raising over eighteen million dollars.

2017 saw an explosion of activity—342 token issuances raised almost $5.4 billion. Sales selling out in increasingly shorter periods of time fuelled the frenzy, and many ill-informed investors rushed to get "in on the action," with the project fundamentals becoming less important.

The U.S. Securities and Exchange Commission put out a statement stating that, "If a digital asset sold to U.S. investors had the characteristics of a security (ownership rights, an income stream, or even the expectation of a profit from the efforts of others), it had to abide by U.S. securities laws".

The ICO surge continued for Blockchain start-ups, as it was an easier and faster way to raise a huge amount of money without sacrificing equity in the company.

Benefits

To the issuer:

- Access to seed funding; much faster; few restrictions than via the venture capital route.
- The opportunity to create new, decentralized business models
- A base of participants incentivized to use and test the service, in a new ecosystem.
- No loss of equity in the project (unless the tokens stipulated ownership sharing).
- The Entrepreneur's dream of creating something great via risk.

To the token holder:

- Access to an innovative idea or service that could impact change.
- Possible gain through an increase in the token's price.
- Participation in a new concept, a role in developing a new technology.

Risks

For the issuer:

- Uncertain regulation (possible post-issue clamp-down, fine, or even sentencing)
- Unstable investment (a sell-off by disgruntled users could affect the token price and the viability of the project)
- Little idea of who the token holders are (unlike shareholders)

For the holder:

- No guarantee the project will get developed. Which has been the norm to date.
- No regulatory protection (investment at risk)
- Often scant information about underlying fundamentals; project phases not completed
- Little transparency on token holding structure

Outlook

As the technology matures and the market gains more experience with the concept, and as investors become more sophisticated, the quality of the tokens and the viability of the business models are likely to improve.

Regulators will most likely pay more attention to token sales to protect investors from flimsy or fraudulent sales.

Meanwhile, new types of business models will continue to emerge, fuelled by a new funding system and operating structure. The infrastructure that supports token sales will also continue to grow, with reputable advisors morphing into the "investment banks" of the sector, and new dedicated platforms increasingly enhancing the user experience.

In October 2018, EOS was the leading Blockchain venture with approximately 4.1 billion U.S. dollars raised through ICO.

Overall, ICO investments are really a big step to take without ramping up on the education and knowledge beforehand. It really comes down to your own situation, your own perspective and the need to actually get involved at all. Soon there will safer routes to take with financial institutions offering portfolios just like mutual funds with different risk levels.

There are all kinds of hype about the crypto projections. Some believe one Bitcoin will go up to $150k in the next three years. Really?

- eBay in 1998 goes public: If you had invested one thousand dollars in the IPO, it would be worth $128,500 today.

- Netflix in 2002 goes public: If you had invested one thousand dollars in the IPO, it would be worth nearly $250,000 today. In 2007 Netflix worth 1.2b and in 2019 worth 20b +

There is some effort required to protect your privacy with Bitcoin. All Bitcoin transactions are stored publicly and permanently on the network, which means anyone can see the balance and transactions of any Bitcoin address but not the identity of the user.

One of the interesting and important capabilities is the ability you will have to exchange one cryptocurrency into another. All this using an application on your phone. "You can now transfer value instantaneously across Blockchains," Litecoin creator Charlie Lee told International Business Times. "This opens up tons of new possibilities including decentralised exchanges."

What you need to understand is Bitcoin is like keeping money with you. And you are responsible for buying, selling, and securely storing it.

Like any financial portfolio, you need balance. Some form of investment in these currencies should only follow a personal education of the biggest change since the internet. The spirit of all crypto is that it will change the world. The bulk of the money in America is owned by people over sixty. Millennials and GenX'ers will not tolerate the current system. Digital money will take root. Everyone needs to understand the fallacies of the current system.

Price Oddities

Bitcoin is ultimately worth what people will buy and sell it for. This is often as much a matter of human psychology as economic calculation.

If your aim is to accumulate Bitcoin, no matter the price dollar-cost averaging will allow you to accumulate Bitcoins at a decent average price, without the stress of trying to predict the sometimes wild gyrations of Bitcoin's price.

Major Downside Risks

Bitcoin is an experimental project and as such, a highly risky asset. There are many negative influencers of price, chief among them being the legislative risk of a major government banning or strictly regulating Bitcoin businesses. The emergence of credible competitors, perhaps with the backing of major (central) banks, could see Bitcoin lose market share in the future.

Back Up Your Wallet

Stored in a safe place, a backup of your wallet can protect you against computer failures and many human mistakes. It can also allow you to recover your wallet after your mobile or computer was stolen if you keep your wallet encrypted.

Encrypt Online Backups

Any backup that is stored online is highly vulnerable to theft. Even a computer that is connected to the Internet is vulnerable to malicious software. As such, encrypting any backup that is exposed to the network is a good security practice. You might also want to consider using different medias like USB keys, papers and CDs.

Encrypting your wallet or your smartphone allows you to set a password for anyone trying to withdraw any funds. This helps protect against thieves, though it cannot protect against keylogging hardware or software. Never forget your password.

You should make sure you never forget the password or your funds will be permanently lost. Unlike your bank, there are very limited password recovery options with Bitcoin. In fact, you should be able to remember your password even after many years without using it. In doubt, you might want to keep a paper copy of your password in a safe place like a vault.

Cryptocurrencies are protected by secret cryptographic keys. Whoever knows those keys can instantly and anonymously move the funds by spending them. This makes theft of cryptocurrency private keys attractive to attackers, who have been able to steal cryptocurrency worth millions of dollars.

Cryptocurrency exchanges should be able to manage private keys correctly and protect consumers from both accidental loss and theft. Several methods of key management have been tried; each method is vulnerable to some extent. A common approach to storing cryptocurrency is cold storage, in which the consumer's cryptographic keys are stored on a device that is not connected to the Internet. The keys are stored in a vault.

A vault is digital, secured storage that can store multiple cryptocurrencies. The vault can be treated like a savings account, providing for withdrawals using various verification methods. Once authenticated, consumers can use their private keys to sign cryptocurrency transactions or conduct payment transactions using the cryptocurrency payment infrastructure.

A vault may be implemented as a cloud-based service or with a user-controlled hardware device (e.g. a smart card or USB device). A vault implemented using a user-controlled device with a secure element (SE) is considered more secure because it is not connected to the Internet. It can also come with a security certification by a third party (e.g. Common Criteria), which prevents having to trust an online service.

Additionally, a cryptocurrency vault can be stored at multiple safe locations disconnected from the Internet. That way, if one vault is compromised, another vault has a record of the data. Online vaults are typically protected using biometric scanner access, 24 x 7 video surveillance, and armed guards.

Vaults not only protect cryptocurrencies against malicious theft, they also help consumers who accidentally misplace their private keys. When consumers lose their private keys, coins become irretrievable forever. Consumers can lose their keys either through a hard drive crash or a software malfunction. The simplest solution to this problem is for consumers to print their private keys (or to use a seed to recover all of them, as proposed by the Bitcoin standard BIP 0032) and store the hard copy.

Open Source

There is a lot of collaboration amongst thousands of people developing, changing modifying applications in a completely transparent way. There is nothing hidden in the code.

Less than a decade old, Bitcoin is worth billions. The cryptocurrency promises to revolutionize global finance by placing control of currency in the hands of users, not nations, and make financial exchanges more transparent, efficient, and democratic. That is the expected end result.

But is Bitcoin really a safe bet? Proponents say the hype around the cryptocurrency is warranted that it will be here in 10 years, including executives at JPMorgan and Goldman Sachs, are increasingly jumping on the Bitcoin (block) train.

On the other hand, skeptics suggest this highly volatile digital currency offers a platform for illicit activity, including money laundering and trafficking of humans and drugs, free from government oversight and regulation.

And, they argue, Bitcoin has no intrinsic value—the price is based on market enthusiasm rather than actual utility.

Having said that, among the countries that are not serviced by Coinbase are some of the largest Bitcoin economies such as China, South Korea, and Japan. Additionally, no Bitcoin users in South America or Africa can use the company's service.

Given the growing number of Bitcoin users in Asia, Africa, and South America who are using other wallet providers, the number of Bitcoin users has to be much higher than Coinbase's thirteen million.

Quantum-Computing Breakthrough?

Enter Quantum Computing, military, and state actors. And now, in September of 2019, Google let news of their QC capability slip.

Thursday, September 19, 2019, Wired Newsletter

Google accidentally made computer science history last week. In recent years the company has been part of an intensifying competition with rivals such as IBM and Intel to develop quantum computers, which promise immense power on some problems by tapping into quantum physics. The search company has attempted to stand out by claiming its prototype quantum processors were close to demonstrating "quantum supremacy," an evocative phrase referring to an experiment in which a quantum computer outperforms a classical one. One of Google's lead researchers predicted the company would reach that milestone back in 2017.

Friday, news slipped out that Google had reached the milestone. The Financial Times drew notice to a draft research paper that had been quietly posted to a NASA website in which Google researchers describe achieving quantum supremacy. Within hours, Democratic presidential candidate Andrew Yang was warning that Google's quantum computers could break encryption, and quantum computing researchers were trying to assure the world that conventional computers and security are not obsolete.

The result doesn't mean quantum computers are ready for practical work.The issue is that quantum computers don't use bits; they use quantum bits or qubits. Qubits give quantum computers the power to exponentiate their brute force attacks, which effectively cancels out the computational hardness that took place with the cryptographic primitive.

Google Claims to Have Achieved "Quantum Supremacy":

Google claims its quantum computer just completed a task in minutes. The same task would've taken a conventional computer 10,000 years.

"The problem their machine solves with astounding speed has been very carefully chosen just for the purpose of demonstrating the quantum computer's superiority," Preskill says. It's unclear how long it will take quantum computers to become commercially useful; breaking encryption—a theorized use for the technology—remains a distant hope. "That's still many years out," says Jonathan Dowling, a professor at Louisiana State University.

Currency is now a means of expression. But if everyone can create a currency, how does it derive value and what does it mean?

What is the difference between currency as an expression of popularity, as an expression of desire, a fad, a brand?

But, why not? I think at some point we're going to see things like that happen.

We're going to have hundreds of altcoins. We're going to have thousands of altcoins. We're going to have hundreds of thousands, and then millions of altcoins.

Now, we can choose currency, and a five-year-old can resurrect Pokemon trading cards in a digital form. Maybe the currency that the five-year-old created has monetary value, maybe it doesn't. Most likely, it doesn't. But some will. We need to get used to a world where we have to judge currency not by who issued it, but by who uses it. Or rather, by how many people use it and what they use it for.

CONCLUSION

Bitcoin has become "the fast and the furious" almost appearing out of nowhere. In just a few years, almost every media headline has reported on the hype, on the wild value swings, and speculative ideas of how high it may go along with who is getting hacked. Comparatively, there is little focus or discussion of how it got here, what it means and what problems it may solve. It's a completely new concept but in 10 years from now it will be just like the web.

Technology, as we all know and see is moving fast. There is a roadmap of centuries of historic monetary changes that have happened right into the digital transformation era. Then, a major breakthrough with the internet. And now, an unsolvable problem being solved by the unknown Satoshi Nakamota paper now allows how currency and assets can be moved in a digital form without the middleman. Why it starts to make more sense why cash is disappearing. Is the next layer of internet technology Blockchain going to be the next generation of Internet?

By 2030, most governments around the world will create or adopt some form of virtual currency. Cryptocurrency is more efficient, provides reduced settlement times, and offers increased traceability.

Financial inclusiveness is the most obvious benefit of cryptocurrencies like Bitcoin. As is already evident today, Bitcoin and block chain enable the unbanked population to get banked, and therefore, get paid. One no longer needs to rely on a centralized institution, such as the government or a bank, to give you permission to open a bank account. You can buy and sell Bitcoin on an open market with access to a smartphone.

In the short term, the impact of government-based cryptocurrency will become an area of experimentation, led mostly by developing nations with unstable economies and weak institutions. Creating a new cryptocurrency presents a viable solution for those governments that need to alleviate the bleak perception of its country's monetary challenges. Venezuela's new cryptocurrency "petro" backed by the oil and gas reserves claimed to raise over $5 billion, many other countries will follow suit.

However, many of these early projects will inevitably fail due to the early stage of the technology which is yet to mature. This situation parallels with the early days of the web in Dot.com era, where major companies developing their e-commerce strategy made mistakes in initial implementations, resulting in huge losses of data and funds. Blockchain's database structure is built as a distributed ledger that ensures complete privacy, security, transparency and control of all data to all users.

Currency as a means of expression, currency as a tool of language, is no longer up to the issuer. It is up to us as individuals, making a choice to use that currency, and we give it value through our use. We give it value through adoption. We will be surprised by some of the currencies that will emerge from a fad, a joke, perhaps even a sick joke, and will explode into viral consciousness on the internet and then become real monetary powers in use across a broad population. We're going to start treating currency as an application, and in order to do that we're going to need interfaces that allow us a unifed currency experience, that allow us to have a single wallet with perhaps 150 different currencies in it. Just like when I choose an application on the internet to communicate with, I'm also aligning myself with a corresponding community. Bitcoin and all the cryptocurrencies will be on a roller coaster for a long time. Traders and investors will make and lose fortunes in record time, betting on them—rolling the dice. There are still many obstacles Bitcoin has to overcome, but the "people's currency" holds a great promise: to become the first true global currency, free of the control of central banks that print money and big banks that generate credit. Or it may provide the foundation and survival for a multitude of other currencies that are building on and from the Bitcoin technology.Bitcoin still needs to build the trust of the "general public."When it does it will become a medium of exchange, standard of value, and store of value, replacing national currencies.

Identity for All

If e-mail and the web proved to be the "killer app" for the Internet, identity solutions will prove to be the "killer app" for blockchain. Blockchain-based identity systems will provide a single source of verification for individuals' identities and assets. Blockchain-based identity decentralizes the data collection, cross-verifies the collected data and stores this information on a decentralized immutable ledger. Over 1.5 billion people in the developing world lack proof of identity, including more than 65 million refugees. The new identity platform will be more secure and reliable since it will be stored on a distributed ledger rather than being in the possession of a central authority. The decision to disclose identity information will be within each individual's control. Blockchain technology has the potential to shrink the poverty gap. How? It can be done by increasing financial inclusiveness, reducing corruption, and enabling decentralized access to value-creating assets. All assets, including land, will be recorded on a transparent, tamper-free distributed ledger open for the public to see.

Disrupting the Supply Chain

By 2030, most of world trade will be conducted leveraging blockchain technology in the global supply chain.

The current state of the global supply chain and world trade is chaotic, fragmented operating in business silos among parties that are untrusted. Counterfeit medicines and the food supply chain are major real world problems and cannot be fully solved without using blockchain technology. Once it matures, the problems in supply chains are real enough, and important enough that solutions will eventually be built, and blockchain will play a critical role in these future solutions.

Just like many different industry applications we have today, we will likely end up with an ecosystem of specialized chains, operating security, privacy controls, transparency and trust in parallel.The impact of Blockchain technology may be greater than any of us realize. There will be the creation of limitless applications and 'new killer apps' to come that can be built on top of it. The potential for change is immense and we're just beginning to scratch the surface of what can be.

Understanding and becoming a little more familiar with the new money should give you comfort level and hopes to create more awareness while having the ability to have a reasonable intelligent conversation. The impact for many of us is more of getting to know the basics of cryptocurrencies without the in-depth technical descriptions. The positive world impact Blockchain will bring for business, governments and most importantly - all people.

There is no necessity to run and buy Bitcoin in a fear of being left behind. At the end of the day, or the end of the decade, Bitcoin and Blockchain will be everywhere with a roadmap on solutions to real world problems.

In 10 years from now, just like the internet, "It just works"!

What more is possible?

ABOUT THE AUTHOR

Ted has more than 10 years of experience in the cybersecurity field. He worked in a trusted consultative role in the cybersecurity field helping Fortune 100 companies Globally, including J.P.Morgan, BNP Parabas, and Humana, to protect their critical data.

He first became aware of Bitcoin in 2012 when his clients were hit by hackers who breached the company firewalls and used a form of malware called Ransomware to encrypt company data. In effect, they could no longer access the data. If the companies wanted the data to return to its normal state they would have to pay a ransom. They had no choice but to follow the hacker's instructions to pay the ransom in Bitcoin. Bitcoin provided the hacker's anonymity and was difficult to track by forensic accountants' police crime units.

While attending the largest cybersecurity trade event in San Francisco, the number of discussions, seminars, and the breakout sessions had become hot topics and the focus was mostly on Bitcoin and Blockchain technology.

Ted has always been passionate about Cybersecurity and the internet of tomorrow. Thus, he decided to convert his years of experience and expertise into a book. Ted decided to write this book to create a new awareness for everyone who has heard about Bitcoin & Blockchain but has no knowledge about it or very little of the benefits and how it came to be. Today, 97% of money transactions are done with digital money. It is moving at warp speed so the attempt here is to provide the basics, the foundation of what it is and what is the future of digital money.

This book is a progressive outlining of things you may not know about the internet, the Dark Web, exponential technology, store of value, historic briefs on how the banks became powerful, and how Blockchain printed form has been used for centuries to what it represents in today's technological form. More importantly, is the thorough understanding of the change that will happen and what next is possible.

Made in the USA
Columbia, SC
09 January 2020

86332338R10089